New Aspects of Quantity Surveying Practice

Third edition

Duncan Cartlidge

Spon Press
an imprint of Taylor & Francis
LONDON AND NEW YORK

First edition published 2002
By Butterworth Heinemann

Second edition published 2006
by Butterworth Heinemann

This edition published 2011
by Spon Press
2 Park Square, Milton Park, Abingdon, Oxon OX14 4RN

Simultaneously published in the USA and Canada
by Spon Press
711 Third Avenue, New York, NY 10017

Spon Press is an imprint of the Taylor & Francis Group, an informa business

British Library Cataloguing in Publication Data
A catalogue record for this book is available from the British Library

Library of Congress Cataloging-in-Publication Data
A catalog record has been requested for this book

ISBN13: 978–0–415–58042–7 (hbk)
ISBN13: 978–0–415–58043–4 (pbk)
ISBN13: 978–0–203–85110–4 (ebk)

Typeset in Sabon by Swales & Willis Ltd, Exeter, Devon

Printed and bound in Great Britain by
CPI Antony Rowe, Chippenham, Wiltshire

To Holly, Lottie and Boris

Contents

Illustrations

Figures

Tables

Foreword

Sustainable development is now high on the agendas of governments, institutions and corporations across the globe. The world's focus on climate change, resource depletion and environmental degradation is strongly influencing the way we deliver and manage the built environment.

We face challenging times in our industry. West striving to engage East, merging and expanding business networks across the globe, seeking out vibrant economies in a time of recession at home, integrating business cultures and challenging our traditional procurement and business ethics. Sustainable development drivers vary in response to national context, requiring international design team members to identify, measure and manage differing and diverse development outcomes against a backdrop of increasingly integrated, complex legislation and regulation.

In many countries, national sustainability initiatives offer performance branding of new and refurbished buildings, integrating design team activities beyond the comfort of linear RIBA Stages A through L, drawing on the power and convenience of advanced, seamless IT design and engineering tools. Quantity surveyors are now challenged with protecting 'value for money' throughout this fast moving, integrated, technology-rich process, while understanding attributes of new and complex materials, assemblies and construction processes, and combinations of passive and active building systems. Cost benefit linked to thermal performance, embodied carbon and durability must be considered in the context of life cycles, buildability, maintainability and functionality.

An immediate challenge to the quantity surveying profession internationally is to develop and implement effective processes and tools, which integrate into the building information modelling environment. This will provide the quantity surveyor with concise, accurate information, at early design stage, allowing well-informed options appraisals against the multitude of design solutions being proposed through complex design and engineering models. Such an approach will require corporate investment in applied research, professional development and infrastructure, while becoming immersed within an ethos of sustainability across the business.

One can only hope that short-term recessionary pressures do not deviate the industry and the profession away from sustainable development.

Francis R. Ives
Chairman 1991 to 2010
Cyril Sweett

Preface to First Edition

The Royal Institution of Chartered Surveyor's Quantity Surveying Think Tank, *Questioning the Future of the Profession*, heard evidence that many within the construction industry thought chartered quantity surveyors were arrogant, friendless and uncooperative. In addition, they were perceived to add nothing to the construction process, failed to offer services which clients expected as standard and too few had the courage to challenge established thinking. In the same year Sir John Egan called the whole future of quantity surveying into question in the Construction Industry Task Force report *Rethinking Construction*, and if this weren't enough a report by the University of Coventry entitled 'Construction Supply Chain Skills Project' concluded that quantity surveyors are 'arrogant and lacking in interpersonal skills'. Little wonder then that the question was asked: 'Will we soon be drying a tear over a grave marked "RIP Quantity Surveying, 1792–2000?" Certainly the changes that have taken place in the construction industry during the past twenty years would have tested the endurance of the most hardy of beasts. Fortunately, the quantity surveyor is a tough and adaptable creature, and to quote and paraphrase Mark Twain, 'reports of the quantity surveyor's death are an exaggeration'.

I have spent the past thirty years or so as a quantity surveyor in private practice, both in the UK and Europe, as well as periods as a lecturer in higher education. During this time I have witnessed a profession in a relentless search for an identity, from quantity surveyor to building economist, to construction economist, to construction cost adviser, to construction consultant, etc. I have also witnessed and been proud to be a member of a profession that has always risen to a challenge and has been capable of reinventing itself and leading from the front, whenever the need arose. The first part of the twenty-first century holds many challenges for the UK construction industry as well as the quantity surveyor, but of all the professions concerned with the procurement of built assets, quantity surveying is the one that has the ability and skill to respond to these challenges.

This book, therefore, is dedicated to the process of transforming the popular perception that, in the cause of self-preservation, the quantity surveyor is wedded to a policy of advocating aggressive price-led tendering with all the problems that this brings, to one of a professional who can help deliver high-value capital projects on time and to budget with guaranteed life cycle costs. In addition, it is hoped that this book will demonstrate beyond any doubt that the quantity surveyor is alive and well, adapting to the demands of construction clients and, what is more, looking forward to a long and productive future. Nevertheless, there is still a long hill to climb. During the production of this book I have heard major construction clients call the construction industry 'very unprofessional' and the role of the quantity surveyor compared to that of a 'post-box'.

In an address to the Royal Institution of Chartered Surveyors in November 2001, the same Sir John Egan who had called the future of the quantity surveyor into question now, as Chairman of the Egan Strategic Forum for Construction, suggested that the future for Chartered Surveyors in construction was to become process integrators, involving themselves in the process management of construction projects, and that those who clung to traditional working practices faced an uncertain future. The author would whole-heartedly agree with these sentiments.

'The quantity surveyor is dead – long live the quantity surveyor – masters of the process!'

www.duncancartlidge.co.uk
Duncan Cartlidge

Preface to the Second Edition

Wanted: quantity surveyors

Four years have passed since the first edition of *New Aspects of Quantity Surveying Practice*. At that time *Building*, the well-known construction industry weekly, described quantity surveying as 'a profession on the brink' while simultaneously forecasting the imminent demise of the quantity surveyor, and references to 'Ethel the Aardvark goes Quantity Surveying' had everyone rolling in the aisles. In a brave new world where confrontation was a thing of the past and where the RICS tried to deny that quantity surveyors existed at all, clearly there was no need for the profession! But wait; what a difference a few years can make, since on 29 October 2004 the same publication that forecast the end of the quantity surveyor had to eat humble pie when the *Building* editorial announced that 'what quantity surveyors have to offer is the height of fashion – Ethel is history'. It would seem as if this came as a surprise to everyone, except quantity surveyors!

Ironically, in 2006 quantity surveyors are facing a very different challenge to the ones that were predicted in the late 1990s. Far from being faced with extinction the problem now is a shortage of quantity surveyors that has reached crisis point, particularly in major cities like London. The 'mother of all recessions between 1990 and 1995' referred to in Chapter 1 had the effect of driving many professionals, including quantity surveyors, out of the industry for good, as well as discouraging school leavers thinking of embarking on surveying degree courses. As a consequence there is now a generation gap in the profession and with the 2012 London Olympics on the horizon, as well as buoyant demand in most property sectors, many organisations are offering incentives and high salaries to attract and retain quantity surveying staff. In today's marketplace a 'thirty-something' quantity surveyor with ten to fifteen years' experience is indeed a rare but not endangered species. It would also seem that the RICS has had second thoughts about the future of the quantity surveyor. A survey carried out by the Royal Bank of Scotland in 2005 indicated that quantity surveyors are the best-paid graduate professionals. In November 2005, the RICS announced that, after years of protest, the title 'quantity surveyor' was to reappear as an RICS faculty as well as the RICS website.

The new millennium found the construction industry and quantity surveying on the verge of a brave new world – an electronic revolution was coming, with wild predictions on the impact that IT systems and electronic commerce would have upon the construction industry and quantity surveying practice; the reality is discussed in Chapter 5.

For the quantity surveyor, the challenges keep on coming. For many years the UK construction industry has flirted with issues such as whole-life costs and sustainability/green issues; it now appears that these topics are being taken more seriously and are discussed in Chapter 3. The RICS Commission on Sustainable Development and Construction recently developed the following mission statement: 'To ensure that sustainability becomes and remains a priority issue throughout the profession and RICS', and committed itself to raising the profile of sustainability through education at all levels from undergraduate courses to the APC. In the public sector, the new Consolidated EU Public Procurement Directive due for implementation in 2006 now makes sustainability a criterion for contract awards and a whole raft of legislation due in spring 2006 has put green issues at the top of the agenda. Links have now been proved between the market value of a building and its green features and related performance.

Following the accounting scandals of the Enron Corporation in 2003 quantity surveyors are being called on to bring back accountability to both the public and private sectors and worldwide expansion of the profession continues with further consolidation and the emergence of large firms moving towards supplying broad business solutions tailored to particular clients and sectors of the market.

Where to next?

Duncan Cartlidge
www.duncancartlidge.co.uk

Preface to Third Edition

Quantity surveying remains a diverse profession with practitioners moving into new areas, some of which are outlined in Chapter 7.

The Preface to the Second Edition of *New Aspects of Quantity Practice* referred to the increasing interest from both the profession and the construction industry in sustainability and green issues. During the past five years, since the previous edition, sustainability has risen to world prominence and the construction industry, worldwide, has been identified as number one in the league table of polluters and users of diminishing natural resources. It is unsurprising therefore that sustainability has risen to prominence in the industry with many undergraduate and postgraduate programmes now including dedicated modules on sustainable development, and clients, professionals, developers and contractors seeking to establish their green credentials.

Ethics, both personal and business, and professional standards have also risen to prominence. Never before has the behaviour of politicians, public figures and professionals been under such close scrutiny; the age of transparency and accountability may truly be said to have arrived. Although ethics has a long history of research and literature in areas such as medicine, the amount of guidance available for surveyors has been almost non-existent until recently and even now cannot be described as comprehensive.

One thing that has been a common theme throughout the writing of the three editions of this book is that quantity surveyors feel unloved not least by their professional institution, the RICS. In 2010 quantity surveyors, not for the first time, threatened to leave the RICS in response to the introduction of AssocRICS, a new grade of membership that, it was believed, would result in a lowering of entry standards to the institution.

Layered on top of the above is what has been described as the deepest recession since the 1920s, with all the challenges that this brought. As this book goes to press, it is still uncertain how many large public sector projects will be axed as the aftermath of the credit crunch lingers on and continues to impact upon the construction industry's order books.

Nevertheless, despite world recessions and new areas of focus for practice, the quantity surveyor continues to prosper, with interest in the profession never higher. Therefore, raise a glass to the quantity surveyor, by any definition a true survivor.

Duncan Cartlidge
www.duncancartlidge.co.uk

Acknowledgements

My thanks go to the following for contributing to this book:

Rohinton Emmanuel BSc (Built Env.); MSc (Architecture); MS (La.St.Uni.); PhD (Michigan); AIA (Sri Lanka); RIBA. Rohinton Emmanuel is a Reader in Sustainable Design and Construction at Glasgow Caledonian University. As an architect with urban design interests, he has pioneered the inquiry of urban climate change in warm regions and has taught and consulted on climate and environment-sensitive design, building energy efficiency, thermal comfort, urban air quality and urban transport planning. Rohinton is the Secretary-elect and an elected member of the Board of the International Association for Urban Climate, and is a member of the Expert Team on Urban and Building Climatology (ET 4.4) of the World Meteorological Organization (WMO) as well as the CIB Working Group (W108) on Buildings and Climate Change. He has also worked as a green building consultant (LEED certification) and has authored over fifty research papers in the areas of climate change in the built environment, building and urban energy efficiency and thermal comfort. Most recently he published *An Urban Approach to Climate Sensitive Design: Strategies for the Tropics* (Taylor & Francis, London, 2005) (Rohinton.emmanuel@gcal.ac.uk).

The BioRegional Development Group for images of the BedZED and Mata de Sesimbra developments.

Mohammed Khirsk, Robert Gordon University, for his contribution to Chapter 3.

John Goodall of FIEC, Brussels.

Chapter 1

The story so far

Introduction

The quantity surveyor has been an integral part of the UK construction industry for around 170 years. The golden age for quantity surveyors may be thought of as the period between 1950 and 1980, when bills of quantities were the preferred basis for tender documentation and the RICS scales of fees were generous and unchallenged. As described in Chapter 1, this situation was due to change beyond all recognition in the latter stages of the twentieth century as a client-led crusade for value for money and leaner project completion times and procurement strategies placed new demands on the construction industry.

The catalyst of change

This chapter examines the root causes of the changes that took place in the United Kingdom construction industry and quantity surveying practice during the latter half of the twentieth century. It sets the scene for the remaining chapters, which go on to describe how quantity surveyors are adapting to new and emerging markets and responding to client-led demands for added value. One of the most important new challenges is how the construction industry and its associated professionals rise to the impact of sustainability and green issues. The construction industry is no stranger to fluctuations in workload, the most recent being the downturn in construction orders following the world financial crisis in 2009 after the collapse of some major financial institutions in America left the world on the brink of a 1920s-style depression. The housing sector was particularly badly affected and the downturn resulted in an 8.2 per cent (67,000) fall in construction-related jobs, the largest of any major industrial sector during this period (Office of National Statistics, 2009). However, as dramatic and concerning as this downturn was, it was the period between 1990 and 1995 that will be remembered, as an eminent politician once remarked, as 'the mother of all recessions'. Certainly, from the perspective of the UK construction industry, this recessionary phase was the catalyst for many of the changes in working practices and attitudes that have been inherited by those who survived this period and continue to work in the industry. As described in the following chapters, some of the pressures for change in the UK construction industry and its professions – including quantity surveying – have their origins in history, while others are the product of the rapid transformation in business practices that took place during the last decades of the twentieth century and still continue today. Perhaps the lessons learned during this period enabled the industry to weather the 2009 financial storm more easily than otherwise would have been the case. This book will therefore examine the background and causes of these changes, and then continue to analyse the consequences and effects on contemporary surveying practice.

Historical overview

The year 1990 was a watershed for the UK construction industry and its associated professions. As illustrated in Figure 1.1, by 1990 a 'heady brew of change' was being concocted on fires fuelled by the recession that was starting to have an impact on the UK construction industry. The main ingredients of this brew, in no particular order, were:

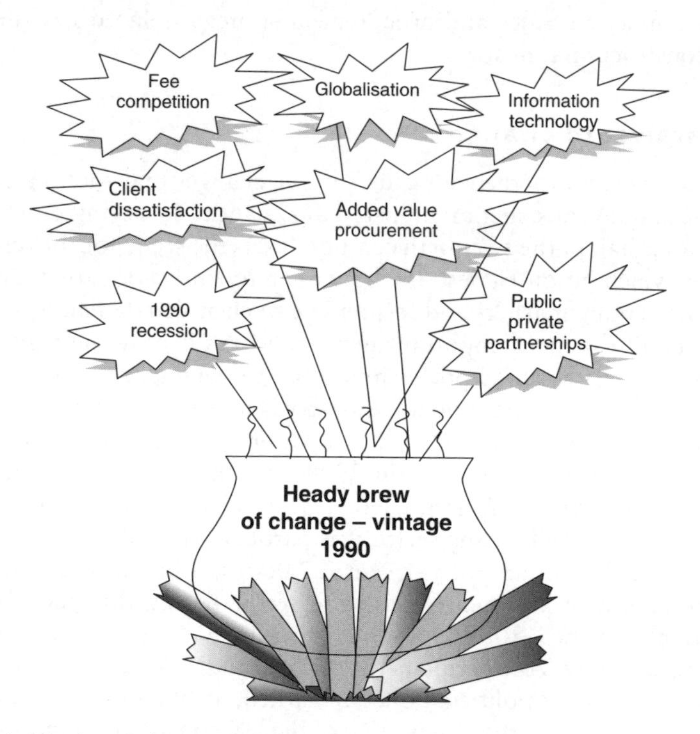

Figure 1.1 The heady brew of change

- The traditional UK hierarchical structure that manifested itself in a litigious, fragmented industry, where contractors and subcontractors were excluded from most of the design decisions.
- Changing patterns of workload due to the introduction of fee competition and compulsory competitive tendering.
- Widespread client dissatisfaction with the finished product.
- The emergence of privatisation and public–private partnerships.
- The pervasive growth of information technology.
- The globalisation of markets and clients.

A fragmented and litigious industry

Boom and bust in the UK construction industry has been and will continue to be a fact of life (see Figure 1.2), and much of the industry, including quantity surveyors, had learned to survive and prosper quite successfully in this climate.

The rules were simple: in the good times a quantity surveyor earned fee income as set out in the Royal Institution of Chartered Surveyors' Scale of Fees for the preparation of, say, Bills of Quantities, and then in the lean times endless months or even years would be devoted to performing countless tedious re-measurements of the same work – once more for a fee. Contractors and subcontractors won work, albeit with very small profit margins, during the good times, and then when work was less plentiful they would turn their attentions to the business of the preparation of claims for extra payments for the inevitable delays and disruptions to the works. The standard forms of contract used by the industry, although heavily criticised by many, provided the impetus (if impetus were needed) to continue operating in this way. Everyone, including the majority of clients, appeared to

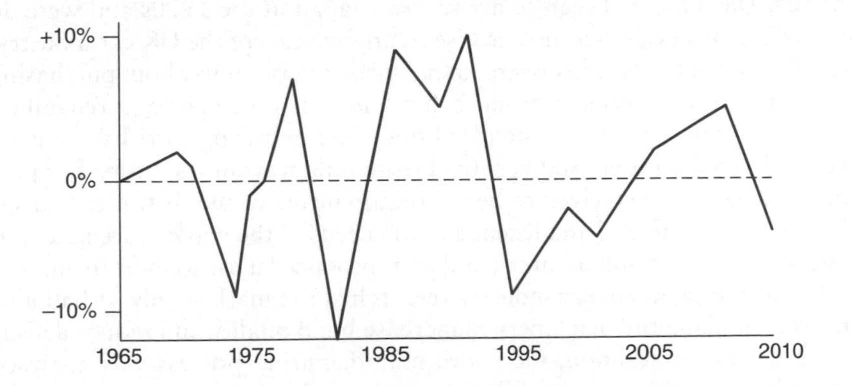

Figure 1.2 Construction output – percentage change 1965–2010

Source: Department.for Business, Innovation and Skills (BIS)

be quite happy with the system, although in practice the UK construction industry was in many ways letting its clients down by producing buildings and other projects that were, in a high percentage of cases, over budget, over time and littered with defects. Time was running out on this system, and by 1990 the hands of the clock were at five minutes before midnight. A survey conducted in the mid-1990s by *Property Week*, a leading property magazine among private sector clients who regularly commissioned new buildings or refurbished existing properties, provided a snapshot of the UK construction industry at that time. In response to the question 'Do projects finish on budget?', 30 per cent of those questioned replied that it was quite usual for projects to exceed the original budget.

In response to the question 'Do projects finish on time?', once more over 30 per cent of those questioned replied that it was common for projects to overrun their planned completion by one or two months. Parallels between the construction industry ethos at the time of this survey and the UK car industry of the 1960s make an interesting comparison. Austin, Morris, Jaguar, Rolls Royce, Lotus and marques such as the Mini and MG were all household names during the 1960s; today they are all either owned by foreign companies or out of business. At the time of writing the first edition of this book Rover/MG, then owned by Phoenix (UK), was the only remaining UK-owned carmaker; now Rover/MG has been confined to the scrap heap when it ceased trading in 2005 amidst bitter recriminations. Rover's decline from being the UK's largest carmaker in the 1960s is a living demonstration of how a country's leading industry can deteriorate, as well as being a stark lesson to the UK construction industry. The reasons behind the collapse of car manufacturing were: flawed design, wrong market positioning, unreliability and poor build quality, but importantly to this may be added lack of investment in new technology, and a failure to move with the times and produce what the market (i.e. the end-users) demanded. Therefore, when the first Datsun cars began to arrive from Japan in the 1970s and were an immediate success, it was no surprise to anyone except the UK car industry. The British car buyer, after overcoming initial reservations about purchasing a foreign car, discovered a product that had nearly 100 per cent reliability, contained many features as standard that were extras on British-built cars, were delivered on time, and benefited from long warranties. Instead of producing what they perceived to be the requirements of the British car buyer, Datsun had researched and listened to the needs of the market, seen the failings of the home manufacturers, and then produced a car to meet them. Not only had the Japanese car industry researched the market fully: it had also invested in plant and machinery to increase build quality and reduce defects in their cars. In addition, the entire manufacturing process was analysed and a lean supply chain established to ensure the maximum economies of production. The scale of the improvements achieved in the car industry are impressive, with the time from completed design to launch reduced from forty to fifteen months, and the supplier defects to five parts per million. So why by 1990 was the UK construction industry staring into the same abyss that the carmakers had faced thirty years earlier? In order to appreciate the situation that existed in the UK construction industry in the pre-1990 period it is necessary to examine the working practices of the UK construction industry, including the role of the contractor and the professions at this time. First, we will take a look back at recent history, and in particular at the events that took place in Europe in the first part of the nineteenth century and which helped to shape UK practice (Goodall, 2000).

The UK construction industry – a brief history

Prior to the Napoleonic Wars, Britain, in common with its continental neighbours, had a construction industry based on separate trades. This system still exists in France as 'lots sépare', and variations of it may be found throughout Europe, including in Germany. The system works like this. Instead of the multi-traded main contractor that operates in the UK, each trade is tendered for and subsequently engaged separately under the coordination of a project manager, or 'pilote'. In France, smaller contractors usually specialise in one or two trades, and it is not uncommon to find a long list of contractors on the site board of a construction project.

The Napoleonic Wars, however, brought change and nowhere more so than in Britain – the only large European state that Napoleon failed to cross or occupy. Paradoxically, the lasting effect which the Napoleonic Wars had on the British construction industry was more profound than on any other national construction industry in Europe.

While it is true that no military action actually took place on British soil, nonetheless the government of the day was obliged to construct barracks to house the huge garrisons of soldiers that were then being transported across the English Channel. As the need for the army barracks was so urgent and the time to prepare drawings, specifications and so on was so short, the contracts were let on a 'settlement by fair valuation based on measurement after completion of the works'. This meant that constructors were given the opportunity and encouragement to innovate and to problem solve – something that was progressively withdrawn from them in the years to come. The same need for haste, coupled with the sheer magnitude of the individual projects, led to many contracts being let to a single builder or group of tradesmen 'contracting in gross', and the general contractor was born. When peace was made the Office of Works and Public Buildings, which had been increasingly concerned with the high cost of measurement and fair value procurement, in particular in the construction of Buckingham Palace and Windsor Castle, decided enough was enough. In 1828, separate trades contracting was discontinued for public works in England in favour of contracting in gross. The following years saw contracting in gross (general contracting) rise to dominate, and with this development the role of the builder as an innovator, problem solver and design team member was stifled to the point where contractors operating in the UK system were reduced to simple executors of the works and instructions (although in Scotland the separate trades system survived until the early 1970s). However, history had another twist; in 1834 architects decided that they wished to divorce themselves from surveyors and establish the Royal Institute of British Architects (RIBA), exclusively for architects. The grounds for this great schism were that architects wished to distance themselves from surveyors and their perceived obnoxious commercial interest in construction. The top-down system that characterises so much of British society was stamped on the construction industry. As with the death of separate trades contracting, the establishment of the RIBA ensured that the UK contractor was once again discouraged from using innovation. The events of 1834 were also responsible for the birth of another UK phenomenon, the quantity surveyor, and for another unique feature of the UK construction industry – post-construction liability.

The ability of a contractor to re-engineer a scheme design in order to produce maximum buildability is a great competitive advantage, particularly on the international scene (see Table 1.1). A system of project insurance that is already widely available on the Continent is starting to make an appearance in the UK; adopting this, the design and execution teams can safely circumvent their professional indemnity insurance and operate as partners under the protective umbrella of a single policy of insurance, thereby allowing the interface of designers and contractors. However, back to history. For the next 150 or so years the UK construction industry continued to develop along the lines outlined above, and consequently by the third quarter of the twentieth century the industry was characterised by powerful professions carrying out work on comparatively generous fee scales, contractors devoid of the capability to analyse and refine design solutions, forms of contract that made the industry one of the most litigious in Europe, and procurement systems based upon competition and selection by lowest price and not value for money. Some within the industry had serious concerns about procurement routes and documentation, the forms of contract in use leading to excess costs, suboptimal building quality and time delays, and the adversarial and conflict-ridden relationships between the various parties. A series of government-sponsored reports (Simon, 1944; Emmerson, 1962; Banwell, 1964) attempted to stimulate debate about construction industry practice, but with little effect.

It was not just the UK construction industry that was obsessed with navel-gazing during the last quarter of the twentieth century; quantity surveyors had also been busy penning numerous reports into the future prospects for their profession. The most notable of these were: *The Future Role of the Chartered Quantity Surveyor* (1983), *Quantity Surveying 2000 – The Future*

Role of the Chartered Quantity Surveyor (1991) and *Challenge for Change: QS Think Tank* (1998), all produced either directly by or on behalf of the Royal Institution of Chartered Surveyors. The 1971 report, *The Future Role of the Quantity Surveyor* (RICS), was the product of a questionnaire sent to all firms in private practice together with a limited number of public sector organisations; sadly, but typically, the survey resulted in a mere 35 per cent response rate. The report paints a picture of a world where the quantity surveyor was primarily a producer of Bills of Quantities; indeed, the report comes to the conclusion that the distinct competence of the quantity surveyor of the 1970s was measurement – a view, it should be added, still shared by many today. In addition, competitive single-stage tendering was the norm, as was the practice of receiving most work via the patronage of an architect. It was a profession where design and construct projects were rare, and quantity surveyors were discouraged from forming multidisciplinary practices and encouraged to adhere to the scale of fee charges. The report observes that clients were becoming more informed, but there was little advice about how quantity surveyors were to meet this challenge. A mere twenty-five years later the 1998 report, *The Challenge for Change*, was drafted in a business climate driven by information technology, where quantities generation is a low-cost activity and the client base is demanding that surveyors demonstrate added value. In particular, medium-sized quantity surveying firms (i.e. between 50 and 250 employees: EU Definition) were singled out by this latest report to be under particular pressure, owing to:

- Competing with large practices' multiple disciplines and a greater specialist knowledge base
- Attracting and retaining a high-quality workforce
- Achieving a return on the necessary investment in IT
- Competing with the small firms with low overheads.

In fact many of small and medium-sized practices that were flourishing in the period 1960 to 1990 have now disappeared victims of mergers and acquisitions of the ever-growing mega-practices and consequently, the surveying profession has been polarised into two groups:

- the large multidisciplinary practices capable of matching the problem-solving capabilities of the large accountancy-based consulting firms;
- small practices that can offer a fast response from a low-cost base for clients, as well as providing services to their big brother practices.

Interestingly, the *Challenge of Change* report also predicts that the distinction between contracting and professional service organisations will blur, a quantum leap from the 1960s, when chartered surveyors were forced to resign from their institution if they worked for contracting organisations!

The trend for mergers and acquisitions continues, although it has to be said not without its problems, with the largest quantity surveying firms developing into providers of broad business solutions.

Since the publication of the final Egan Report in 2002 there have been no more major studies into the UK construction industry; either we are all suffering from report fatigue or we have come to the conclusion that for the present UK construction is at last getting matters more or less right at last. Even so there is no room for complacency.

Measuring performance

Benchmarking

Benchmarking is a generic management technique that is used to compare performance between varieties of strategically important performance criteria. These criteria can exist between different organisations or within a single organisation provided that the task being compared is a similar process. It is an external focus on internal activities functions or operations aimed at achieving continuous improvement (Leibfried and McNair, 1994). Owing to the diversity of its products and processes, construction is one of the last industries to embrace objective performance measurements. There is a consensus among industry experts that one of the principal barriers to promoting improvement in construction projects is the lack of appropriate performance measurement and this was referred to also in Chapter 3 in relation to whole life costs calculations. For continuous improvement to occur it is necessary to have performance measures which check and monitor performance, to verify changes and the effect of improvement actions, to understand the variability of the process, and in general it is necessary to have objective information available in order to make effective decisions. Despite the late entry of benchmarking to construction this does not diminish the potential benefits that could be derived; however, it gives some indication of the fact that there is still considerable work to be undertaken both to define the areas where benchmarking might be valuable and the methods of measurement. The current benchmarking and KPI programme in the UK construction industry has been headlined as a way to improve underperformance. However, despite the production of several sets of KPIs large-scale improvement still remains as elusive as ever. Why is this?

The Xerox Corporation in America is considered to be the pioneer of benchmarking. In the late 1970s Xerox realised that it was on the verge of a crisis when Japanese companies were marketing photocopiers cheaper than it cost Xerox to manufacture a similar product. It is claimed that by benchmarking Xerox against Japanese companies it was able to improve their market position and the company has used the technique ever since to promote continuous improvement. Yet again, another strong advocate of benchmarking is the automotive industry, which successfully employed the technique to reduce manufacturing faults. Four types of benchmarking may be broadly defined: Internal, Competitive, Functional and Generic (Lema and Price, 1995). However, Carr and Winch (1998), while regarding these categories as important suggest that a more useful distinction in terms of methodology is that of output and process benchmarking. Interestingly, Winch (1996) discovered that the results from a benchmarking exercise could sometimes be surprising, as illustrated in Table 1.1, which shows the performance of the Channel Tunnel project in relation to other multi-million-dollar mega-infrastructure projects throughout the world using benchmarks established by the RAND corporation. The results are surprising because the Channel Tunnel is regularly cited as an example of just how bad the UK construction industry is at delivering prestige projects. By contrast the Winch benchmarking exercise demonstrated that the Channel Tunnel project faired better than average when measured against a range of performance criteria.

Since the Winch study in 1996 there has been a tailing off of traffic using the Channel Tunnel and it is now obvious that projections of growth of users were grossly optimistic.

Measuring performance

Through the implementation of performance measures (what to measure) and selection of measuring tools (how to measure), an organisation or a market sector communicates to the outside world and to clients the priorities, objectives and values to which the organisation or market sector aspires. Therefore the selection of appropriate measurement parameters and procedures is very important to the integrity of the system.

It is now important to distinguish between benchmarks and benchmarking. It is true to say that most organisations which participate in the production of Key Performance Indicators (KPIs) for the Construction Best Practice Programme (CBPP) have to date produced benchmarks. Since the late 1990s there has been a widespread government-backed campaign to intro-

Table 1.1 Results from a benchmarking exercise

Performance criterion	*Mega projects average*	*Channel Tunnel*
Budget increase	88%	69%
Programme overrun	17%	14.2%
Conformance overrun	53% performance not up to expectations	As expected
Operational profitability	72% unprofitable	Operationally profitable but overwhelmed by finance charges

duce benchmarking into the construction industry with the use of so-called Key Performance Indicators (KPIs). The objectives of the benchmarking as defined by the Office of Government Commerce are illustrated in Figure 1.3. Benchmarks provide an indication of position relative to what is considered optimum practice and hence indicate a goal to be obtained, but while useful for giving a general idea of areas requiring performance improvement, they provide no indication of the mechanisms by which increased performance may be brought about. Basically it tells us that we are underperforming but it does not give us the basis for the underperformance. The production of KPIs which has been the focus of construction industry initiatives to date has therefore been concentrated on output benchmarks. A much more beneficial approach to measurement is process benchmarking described by Pickerell and Garnett (1997) 'as analyzing why your current performance is what it is, by examining the process your business goes through in comparison to other organisations that are doing better and then implementing the improvements to boost performance'. The danger with the current enthusiasm in the construction industry for KPIs is that outputs will be measured and presented but processes will not be improved as the underlining causes will not be understood. According to Carr and Winch (1998) many recent benchmarking initiatives in the construction industry have shown that while the principles have been understood and there is much discussion about its potential, 'no one is actually doing the real thing'. Benchmarking projects have tended to remain as strategic goals at the level of senior management.

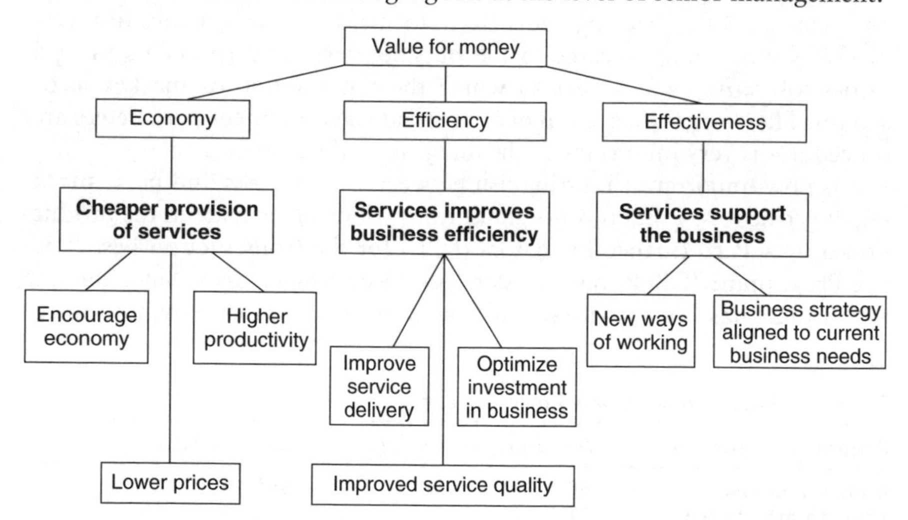

Figure 1.3 Objectives of benchmarking

The performance measures selected by the Construction Industry Best Practice Programme are as follows.

Project performance

- Client satisfaction – product
- Client satisfaction – service
- Defects
- Predictability – cost
- Predictability – time
- Construction – cost
- Construction – time.

Company performance

- Profitability
- Productivity
- Safety
- Respect for people
- Environmental.

These are measured at five key stages throughout the lifetime of a project. The measurement tools range from crude scoring on a 1 to 10 basis to the number of reportable accidents per 100,000 employees. For benchmarking purposes the construction industry is broken down into sectors such as public housing and repair and maintenance.

The British system compared

The following studies offer opportunities to directly compare the British system of procurement and project management with that of a European neighbour: France.

- In the mid 1990s Graham Winch and Andrew Edkins carried out a study on the construction of two identical buildings needed to house a security scanning system as part the Eurotunnel project. A leading UK architectural practice was commissioned for the design on both sides of the Channel, who in turn procured medium-sized British and French firms for the construction. The resultant projects offered a unique opportunity to compare project performance in the two countries with a functionally equivalent building, a common design and a single client. The final analysis demonstrated how the French performed much better than the British in terms of out-turn costs and completion times, despite the fact that both project teams faced similar challenges, largely generated by problems with scanning technology, yet the French team coped with them more smoothly. Why was this?

The answer would seem to lie in the differences in the organisation of the two projects:

- The French contract included detail design, the norm in France; the British contractor was deemed not capable of entering into a design-and-build contract due to the requirements for design information under the JCT form of contract.
- The French contractor re-engineered the project, simplifying the design and taking out unnecessary costs. This was possible owing to the single point project liability that operates in France.
- Under the French contract, the British architect could not object to these contractor-led changes. Under the JCT contract, professional indemnity considerations meant that the architect refused to allow the British contractor to copy the French changes.
- The simplified French design was easier, cheaper and quicker to build. This meant that there was room for manoeuvre as the client-induced variations mounted, whereas the British-run project could only cope by increasing the programme and budget. Once the project began to run late, work on construction became even less effective as the team had to start working out of sequence around the installation of scanning equipment. The researcher's conclusion was that British procurement arrangements tend to generate complexity in project organisation, while the engineering capabilities of French contractors mean that they are able to simplify the design. Indeed, they argue that it is these capabilities that are essential to the French contractor's ability to win contracts.

A second comparison in approaches to construction design and procurement was published in 2004 by the Building Design Partnership, entitled *Learning from French Hospital Design*. Given the massive hospital building programme in the UK, which is planned to continue until at least 2012, the study compared French hospitals with newly built UK hospitals not only from the point of view of design quality but also value for money. The results of the study are given in Table 1.2.

Heath warning: when interpreting the cost data above, it should be remembered that direct comparison of cross-border cost information is notoriously difficult due to a range of factors including building regulations and other statutory controls. Even so:

- French hospitals cost between half and two-thirds of UK hospitals per m^2 but per bed they are more or less similar. Area per bed however is much higher in France, with single bed wards used universally. The report therefore argues that French bed space outperforms its UK counterparts.
- Building service costs in France (i.e. mechanical and electrical installation) are less than half those of the UK, with French comments that the UK overspecifies. More ambitious automation and ICT are also used in France.

Table 1.2 A comparison of UK and French examples

UK examples	*Floor area* m^2	*Total Cost* Euro	*Building Cost* Euro/m^2
Macclesfield	3,353	7,182,634	2,142
Hillingdon	3,600	5,495,717	1,527
Warley	8,940	17,853,354	2,103
Halton, Runcorn	5,493	7,698,900	1,402
French examples			
Montreuil sur Mer	19,691	16,776,184	852
St. Chamond	6,953	11,897,767	1,171
Armberieu	10,551	9,202,711	872
Chateauroux	4,994	6,726,183	1,347

Source: Building Design Partnership (2004).

- Contractor-led detail design seems to lie behind much of the economy of means; many Egan-advocated processes are used. Interestingly consultants' fees, compared to the UK, are high as a percentage of cost.
- In spite of the fact that labour and material costs are higher in France than in the UK, although concrete, France's main structural material is 75 per cent of the UK cost, and out-turn costs over a range of building types, not simply hospitals, are cheaper in France. However, data released by Gardiner and Theobald seem to indicate that recent trends, due in part to the differentials between the British pound and the euro, have seen the gap close.
- The design quality of French hospitals is generally high, while in the UK standards achieved recently have been disappointing and have attracted some criticism.

Compared with many European countries, UK construction produces high output costs for customers from low input costs of professional advice, trade labour and materials. This fact is at the root of the Egan critique, pointing out that the UK has a wasteful system which would cost even more if UK labour rates were equal to those found in Europe. The waste in the system, ten years-plus from Latham, is still estimated to be around 30 per cent. Looking at French design and construction it is possible to see several of the Egan goals in place, but in ways specific to France. While the design process begins with no contractor involvement, contractors become involved sooner than in the UK and take responsibility for much of the detailed design and specification. They are more likely to buy standard components and systems from regular suppliers with well-developed supply chains, rather than on a project-by-project basis. Constructional simplicity follows from the French approach with French architects having little control of details and not appearing to worry too much about door and window details, for example. In the case of French hospitals, despite the lower cost, the projects contain very sophisticated technology with ICT systems becoming very ambitious.

Therefore a simple cost comparison demonstrates that French hospital out-turn costs are cheaper than in the UK, but what of added value? Health outcomes in France are generally superior to those in the UK due to factors such as bed utilisation and patient recovery times, and single rooms instead of multi-patient wards prevent the spread of dangerous so-called 'super-bugs'.

Keeping the focus on Europe, for many observers the question of single point or project liability – the norm in many countries, such as Belgium and France – is pivotal in the search for adding value to the UK construction product, and is at the heart of the other construction industries' abilities to re-engineer designs. Single point project liability insurance protects all the parties involved in both the design and construction process against failures in both design and construction of the works for the duration of the policy. The current system, where some team members are insured and some are not, results in a tendency to design defensively, caveat all statements and advice with exclusions of liability, and not to seek help from other members of the team – not a recipe for team work. In the case of a construction management contract, the current approach to latent defect liability can result in the issue of between twenty and thirty collateral warranties, which facilitates the creation of a contractual relationship where one would otherwise not exist in order that the wronged party is then able to sue under contract rather than rely on the tort of negligence. Therefore, in order to give contractors the power truly to innovate and to use techniques like value engineering (see Chapter 6), there has to be a fundamental change in the approach to liability. Contract forms could be amended to allow the contractor to modify the technical design prior to construction, with the consulting architects and engineers waiving their rights to interfere.

If this approach is an option, why does the UK construction industry still fail to produce the goods? The country is currently holding its breath to see whether the stadia and infrastructure to the 2012 Olympic Games will be completed on time and to budget. The principal problems behind the failure of many high-profile projects were no business case, little or no understanding of the needs of the client, and the inability of a contractor to re-engineer the proposals and produce alternatives. The result: grandiose designs with large price tags and a complete disregard for the need to pay back the cost of the project from revenues generated by the built asset, in this case a sports stadium.

Opponents of the proposal to introduce single-point liability cite additional costs as a negative factor. However, indicative costs given by Royal & SunAlliance seem to prove that these are minimal – for example, traditional structural and weatherproofing: 0.65–1.00 per cent of contract value total cover, including structural, weatherproofing, non-structural and mechanical and electrical; 1–2 per cent of contract value to cover latent defects for periods of up to twelve years, to tie in with the limitations provisions of contracts under Seal. As in the French system, technical auditors can be appointed to minimise risk and, some may argue, add value through an independent overview of the project.

Changing patterns of workload

The patterns of workload that quantity surveyors had become familiar with were also due to change. The change came chiefly from two sources:

1. Fee competition and compulsory competitive tendering (CCT).
2. The emergence of a new type of construction client.

Fee competition and compulsory competitive tendering

Until the early 1970s, fee competition between professional practices was almost unheard of. All the professional bodies published scales of fees, and competition was vigorously discouraged on the basis that a client engaging an architect, engineer or surveyor should base his or her judgement on the type of service and not on the level of fees. Consequently, all professionals within a specific discipline quoted the same fee. However, things were to change with the election of the Conservative government in 1979. The new government introduced fee competition into the public sector by way of its compulsory competitive tendering programme (CCT), and for the first time professional practices had to compete for work in the same manner as contractors or subcontractors – i.e. they would be selected by competition, mainly on the basis of price. The usual procedure was to submit a bid based upon scale of fees minus a percentage. Initially these percentage reductions were a token 5 or 10 per cent, but as work became difficult to find in the early 1980s, practices offered 30 or even 40 per cent reduction on fee scales. It has been suggested that during the 1980s fee income from some of the more traditional quantity surveying services was cut by 60 per cent. Once introduced there was no going back, and soon the private sector began to demand the same reduction in fee scales; within a few years the cosy status quo that had existed and enabled private practices to prosper had gone. The Monopolies and Mergers Commission's (1977) report into scales of fees for surveyors' services led the Royal Institution of Chartered Surveyors to revise its by-laws in 1983 to reduce the influence of fee scales to the level of 'providing guidance' – the gravy train had hit the buffers!

By-law 24 was altered from:

> No member shall with the object of securing instructions or supplanting another member of the surveying profession, knowingly attempt to compete on the basis of fees and commissions

to

> No member shall . . . quote a fee for professional services without having received information to enable the member to assess the nature and scope of the services required.

With the introduction of fee competition the average fee for quantity surveying services (expressed as a percentage of construction cost) over a range of new-build projects was just 1.7 per cent! As a result, professional practices found it increasingly difficult to offer the same range of services and manning levels on such a reduced fee income; they had to radically alter the way they operated, or go out of business. However, help was at hand for the hard-pressed practitioner; the difficulties of trying to manage a practice on reduced fee scale income during the latter part of the 1980s were mitigated by a property boom, which was triggered in part by a series of government-engineered events that combined to unleash a feeding frenzy of property development. In 1988 construction orders peaked at £26.3 billion, and the flames under the heady brew of change were dampened down, albeit only for a few years. The most notable of these events were:

- The so-called Stock Exchange 'Big Bang' of 1986, which had the direct effect of stimulating the demand for high-tech offices.
- The deregulation of money markets in the early 1980s, which allowed UK banks for the first time to transfer money freely out of the country, and foreign finance houses and banks to lend freely on the UK market and invest in UK real estate.
- The announcement by the Chancellor of the Exchequer, Nigel Lawson, of the abolition of double tax mortgage relief for domestic dwellings in 1987, which triggered an unprecedented demand for residential accommodation; the result was a massive increase in lending to finance this sector, as well as spiralling prices and land values.
- Last but by no means least, the relaxation of planning controls, which left the way open for the development of out-of-town shopping centres and business parks.

However, most property development requires credit, and the boom in development during the late 1980s could not have taken place without financial backing. By the time the hard landing came in 1990, many high street banks with a reputation for prudence found themselves dangerously exposed to high-risk real estate projects. During the late 1980s, virtually overnight the banks changed from conservative risk managers to target-driven loan sellers, and by 1990 they found themselves with a total property-related debt of £500 billion. The phenomenon was not just confined to the UK. In France, for example, one bank alone, Credit Lyonnais, was left with 10 billion euros of unsecured loss after property deals on which the bank had lent money collapsed owing to oversupply and a lack in demand; only a piece of creative accountancy and state intervention saved the French bank from insolvency. The property market crash in the early 1990s occurred mainly because investors suffered a lack of confidence in the ability of real estate to provide a good return on investment in the short to medium term in the light of high interest rates, even higher mortgage rates, and an inflation rate that doubled within two years. In part it was also brought about by greed in the knowledge that property values had historically seldom delivered negative values. As large as these sums seem they pale into insignificance when compared to the debts rung up by banks like the Royal Bank of Scotland in the period 2005 to 2008, which reported a £28 billion loss in January 2009 and was only saved from insolvency by a government-led bailout.

The emergence of a new type of construction client

Another vital ingredient in the brew of change was the emergence of a new type of construction client. Building and civil engineering works have traditionally been commissioned by either public or private sector clients. The public sector has been a large and important client for the UK construction industry and its professions. Most government bodies and public authorities would compile lists or 'panels' of approved quantity surveyors and contractors for the construction of hospitals, roads and bridges, social housing and so on, and inclusion on these panels ensured that they received a constant and reliable stream of work. However, during the 1980s the divide between public and private sectors was to blur. The Conservative government of 1979 embarked upon an energetic and extensive campaign for the privatisation of the public sector that culminated in the introduction of the Private Finance Initiative in 1992. Within a comparatively short period there was a shift from a system dominated by the public sector to one where the private sector was growing in importance. Despite this shift to the private sector the public sector remains, for the moment, influential; in 2008, for example, it accounted for 37 per cent or £30 billion of the UK civil engineering and construction industry's business, with a government pledge to maintain this level of expenditure. Nevertheless, the privatisation of the traditional public sector resulted in the emergence of major private sector clients such as the British Airports Authority, privatised in 1987, with an appetite for change and innovation. This new breed of client was, as the RICS had predicted in its 1971 report on the future of quantity surveying, becoming more knowledgeable about the construction process, and such clients were not prepared to sit on their hands while the UK construction industry continued to underperform. Clients such as Sir John Egan, who in July 2001 was appointed Chairman of the Strategic Forum of Construction, became major players in the drive for value for money. The poor performance of the construction industry in the private sector has already been examined; however, if anything, performance in the public sector paints an even more depressing picture. This performance was scrutinised by the National Audit Office (NAO) in 2001 in its report *Modernising Construction* (Auditor General, 2001), which found that the vast majority of projects were over budget and delivered late. So dire has been the experience of some public sector clients – for example, the Ministry of Defence – that new client-driven initiatives for procurement have been introduced. In particular there were a number of high-profile public projects disasters such as the new Scottish Parliament in Edinburgh, let on a management contracting basis which rose in cost from approximately £100 million to £450 million and was delivered in 2004 – two years late and with a total disregard for life cycle costs.

If supply chain communications were polarised and fragmented in the private sector, then those in the public sector were even more so. A series of high-profile cases in the 1970s, in which influential public officials were found to have been guilty of awarding construction contracts to a favoured few in return for bribes, instilled paranoia in the public sector, which led to it distancing itself from contractors, subcontractors and suppliers – in effect from the whole supply chain. At the extreme end of the spectrum this manifested itself in public sector professionals refusing to accept even a diary, calendar or a modest drink from a contractor in case it was interpreted as an inducement to show bias. In the cause of appearing to be fair, impartial and prudent with public funds, most public contracts were awarded as a result of competition between a long list of contractors on the basis of the lowest price. The 2001 National Audit Office report suggests that the emphasis on selecting the lowest price is a significant contributory factor to the tendency towards adversarial relationships. Attempting to win contracts under the 'lowest price wins' mentality leads firms to price work unrealistically low and then seek to recoup their profit margins through contract variations arising from, for example, design changes and other claims leading to disputes and litigation. In an attempt to eradicate inefficiencies the public sector commissioned a number of studies such as the Levene Efficiency Scrutiny in 1995, which recommended that departments in the public sector should:

- Communicate better with contractors to reduce conflict and disputes.
- Increase the training that their staff receive in procurement and risk management.
- Establish a single point for the construction industry to resolve problems common to a number of departments. The lack of such a

management tool was identified as one of the primary contributors to problems with the British Library project.

In June 1997 it was announced that compulsory competitive tendering would be replaced with a system of best value in order to introduce, in the words of the local government minister Hilary Armstrong, 'an efficient, imaginative and realistic system of public sector procurement'. Legislation was passed in 1999, and from 1 April 2000 it became the statutory duty of the public sector to obtain best value. Best value will be discussed in more detail in Chapter 6.

In 2002 the Office of Government Commerce announced that the preferred methods of procurement for the UK public sector would be:

- public–private partnerships
- prime contracting
- design and build.

The information technology revolution

As measurers and information managers, quantity surveyors have been greatly affected by the information technology revolution. Substantial parts of the chapters which follow are devoted to the influence that IT has had and will continue to have, both directly and indirectly, on the quantity surveying profession. However, this opening chapter would not be complete without a brief mention of the contribution of IT to the heady brew of change. To date, mainly individual IT packages have been used or adapted for use by the quantity surveyor – for example, spreadsheets. However, the next few years will see the development of IT packages designed specifically for tasks such as measurement and quantification, which will fundamentally change working practices. The speed of development has been breathtaking. In 1981 the Department of the Environment developed and used a computer-aided bill of quantities production package called 'Enviro'. This then state-of-the-art system required the quantity surveyor to code each measured item, and on completion the codes were sent to Hastings on the south coast of England, where a team of operators would input the codes, with varying degrees of accuracy, into a mainframe computer. After the return of the draft bill of quantities to the measurer for checking, the final document was then printed, which in most cases was four weeks after the last dimensions were taken off!

In recent years architects have made increasing use of computer-aided design (CAD) in the form of 2D drafting and 3D modelling for the production of project information. A report by the Construction Industry Computing Association (2000) entitled *Architectural IT Usage and Training Requirements* indicated that in architectural practices with more than six staff, between 95 and 100 per cent of all those questioned used 2D drafting to produce information. This shift from hand-drawn drafting to IT-based systems has allowed packages to be developed that link the production of drawings and other information to their measurement and quantification, thereby revolutionising the once labour-intensive bill of quantities preparation procedure. Added to this, the spread of the digital economy means that drawings and other project information can be produced, modified and transferred globally. One of the principal reasons for quantity surveyors' emergence as independent professionals during the Napoleonic Wars and their subsequent growth to hold a pivotal role in the construction process had, by the end of the 1990s, been reduced to a low-cost IT operation.

Those who mourn the demise of traditional methods of bill of quantities production should at least take heart that no longer will the senior partner be able to include those immortal lines in a speech at the annual Christmas office party: 'You know, after twenty years of marriage my wife thinks that quantity surveying is all about taking off and working up' – pause for laughter!

As mentioned previously, there had been serious concern both in the industry and in government about the public image of UK Construction plc. The 1990 recession had opened the wounds in the construction industry and shown its vulnerability to market pressures. Between 1990 and 1992 over 3,800 construction enterprises became insolvent, taking with them skills that would be badly needed in the future. The professions also suffered a similar haemorrhage of skills as the value of construction output fell by double-digit figures year on year. The recession merely highlighted what had been apparent for years: the UK construction industry and its professional advisers had to change. The heady brew of change was now complete, but concerns over whether or not the patient realised the seriousness of the situation still gave grounds for concern. The message was clear: industry and quantity surveying must change or, like the dinosaur, be confined to history!

Response to change

In traditional manner, the UK construction industry turned to a report to try to solve its problems. In 1993 Sir Michael Latham, an academic and politician, was tasked to prepare yet another review, this time of the procurement and contractual arrangements in the United Kingdom construction industry. In July 1994, *Constructing The Team* (or the Latham Report, as it became known) was published. The aims of the initiative were to reduce conflict and litigation, as well as to improve the industry's productivity and competitiveness. The construction industry held its breath – was this just another Banwell or Simon to be consigned, after a respectful period, to gather dust on the shelf? Thankfully not. The UK construction industry was at the time of publication in such a fragile state that the report could not be ignored.

This is not to say that it was greeted with open arms by everyone – indeed, the preliminary report, *Trust and Money*, produced in December 1993, provoked profound disagreement in the industry and allied professions.

Latham's report found that the industry required a good dose of medicine, which the author contended should be taken in its entirety if there was to be any hope of a revival in its fortunes. The Latham Report highlighted the following areas as requiring particular attention to assist UK construction industries to become and be seen as internationally competitive:

- Better performance and productivity, to be achieved by using adjudication as the normal method of dispute resolution, the adoption of a modern contract, better training, better tender evaluation, and the revision of post-construction liabilities to be more in line with, say, France or Spain, where all parties and not just the architect are considered to be competent players and all of them therefore made liable for non-performance for up to ten years.
- The establishment of well-managed and efficient supply chains and partnering agreements.
- Standardisation of design and components, and the integration of design, fabrication and assembly to achieve better buildability and functionality.
- The development of transparent systems to measure performance and productivity both within an organisation and with competitors.
- Team work and a belief that every member of the construction team from client to subcontractors should work together to produce a product of which everyone can be justifiably proud.

The Latham Report placed much of the responsibility for change on clients in both the public and private sectors. For the construction industry, Latham set the target of a 30 per cent real cost reduction by the year 2000, a figure based on the CRINE (Cost Reduction Initiative for the New Era) review carried out in the oil and gas industries a few years previously (CRINE, 1994). The CRINE review was instigated in 1992, with the direct purpose of identifying methods by which to reduce the high costs in the North Sea oil and gas industry. It involved a group of operators and contractors working together to investigate the cause for such high costs in the industry, and also to produce recommendations to aid the remedy of such. The leading aim of the initiative was to reduce development and production costs by 30 per cent, this being achieved through recommendations such as the use of standard equipment, simplifying and clarifying contract language, removing adversarial clauses, rationalisation of regulations, and the improvement of credibility and quality qualifications. It was recommended that the operators and contractors work more closely, pooling information and knowledge, to help drive down the increasing costs of hydrocarbon products and thus indirectly promote partnering and alliancing procurement strategies (see Chapter 6). The CRINE initiative recommendations were accepted by the oil and gas industries, and it is now widely acknowledged that without the use of partnering/alliancing a great number of new developments in the North Sea would not have been possible. Shell UK Exploration and Production reported that the performance of the partners in the North Fields Unit during the period 1991 to 1995 resulted in an increase in productivity of 25 per cent, a reduction in overall maintenance costs of 31 per cent in real terms, and a reduction in platform 'down-time' of 24 per cent. Could these dramatic statistics be replicated in the construction industry? 'C' is not only for construction but also for conservative, and many sectors of the construction industry considered 30 per cent to be an unrealistically high and unreachable target. Nevertheless, certain influential sections of the industry, including Sir John Egan and BAA, accepted the challenge and went further, declaring that 50 per cent or even 60 per cent savings were achievable. It was the start of the client-led crusade for value for money.

The Latham Report spawned a number of task groups to investigate further the points raised in the main report, and in October 1997, as a direct result of one of these groups, Sir John Egan, a keen advocate of Sir Michael Latham's report and known to be a person convinced of the need for change within the industry, was appointed head of the Construction Task Force. One of the Task Force's first actions was to visit the Nissan UK car plant in Sunderland to study the company's supply chain management techniques and to determine whether they could be used in construction (see Chapter 6). In June 1998 the Task Force published the report *Rethinking Construction* (DoE, 1998), which was seen as the blueprint for the modernisation of the systems used in the UK construction industry to procure work. As a starting point, *Rethinking Construction* revealed that in a survey of major UK property clients, many continued to be dissatisfied with both contractors' and consultants' performance. Added to the now familiar concerns about failure to keep within agreed budgets and completion schedules, clients revealed that:

- More than one-third of them thought that consultants were lacking in providing a speedy and reliable service.
- They felt they were not receiving good value for money insofar as construction projects did not meet their functional needs and had high whole-life costs.
- They felt that design and construction should be integrated in order to deliver added value.

Frustrated by the lack of change in the construction industry Egan's last act before moving on from the Task Force in 2002 was to pen his final report *Accelerating Change*.

As for quantity surveyors, the 1990s ended with perhaps the unkindness cut of all. The RICS, in its Agenda for Change initiative, replaced its traditional divisions (which included the Quantity Surveying Division) with sixteen faculties, not unlike the system operated by Organisme Professionel de Qualification Technique des Economistes et Coordonnateurs de la Construction (OPQTECC), the body responsible for the regulation of the equivalent of the quantity surveyor in France. It seemed to some that the absence of a quantity surveying faculty would result in the marginalisation of the profession; however, the plan was implemented in 2000, with the Construction Faculty being identified as the new home for the quantity surveyor within the RICS. This move however was not taken lying down by the profession; disillusioned quantity surveyors threatened the RICS with legal action to reverse the decision and in 2004 the Builder Group began to publish a new weekly magazine for quantity surveyors, *QS News*. In 2004 the QSi was formed by Roger Knowles as 'the only professional body that caters solely for quantity surveyors'; although the QSi appears still to be open for business there is no information about the numbers of disillusioned quantity surveyors it has attracted, although the website indicates that the North West Branch of QSi currently has five members.

By 2005 it appeared that the RICS had had a change of mind, with references to quantity surveyors reappearing on the RICS website and rumours of a restructuring of the faculties. Ultimately the existing RICS Faculties were organised into seventeen Professional Groups, one of which is the Built Environment Group where quantity surveying and construction now has its home. So would peace break out between the RICS and its quantity surveying members? The answer is, unfortunately, no.

In 2010 a number of quantity surveying members protested against disenfranchisement and a lowering of entry standards, the latter comment referring to the introduction of a new form of membership AssocRICS to replace the existing TechRICS, an initiative that never really caught the imagination of members. It is feared by the rebellious members that the entry requirements for AssocTech will lower entry standards. In addition, the RICS took the decision to leave the Construction Industry Council, thereby eliminating a uniform voice for construction. Once again, in shades of 2000, quantity surveyors threatened to leave the RICS with claims of being treated as the Cinderellas of the organisation. Isn't there another fairy story about a boy who cried wolf too often?

Beyond the rhetoric

How are the construction industry and the quantity surveyor rising to the challenges outlined above? When the much-respected quantity surveyors Arthur J. and Christopher J. Willis penned the Foreword to the eighth edition of their famous book *Practice and Procedure for the Quantity Surveyor* in 1979, the world was a far less complicated place. Diversification into new fields for quantity surveyors included heavy engineering, coal-mining and 'working abroad'. In the Willis's book, the world of the quantity surveyor was portrayed as a mainly technical back office operation providing a limited range of services where, in the days before compulsory competitive tendering and fee competition, 'professional services were not sold like cans of beans in a supermarket'. The world of the Willises was typically organised around the production of bills of quantities and final accounts, with professional offices being divided into pre- and post-contract services. This model was uniformly distributed across small and large practices, the main difference being that the larger practices would tend to get the larger contracts and the smaller practices the smaller contracts. This state of affairs had its advantages, as most qualified quantity surveyors could walk into practically any office and start work immediately; the main distinguishing feature between practices A and B was usually only slight differences in the format of taking-off paper. However, owing to the changes that have taken place not only within the profession and the construction industry but on the larger world stage (some of which have been outlined in this chapter), the world of the Willises has, like the British car industry, all but disappeared for ever.

In the early part of the twenty-first century, the range of activities and sectors where the quantity surveyor is active is becoming more and more diverse. The small practice concentrating on traditional pre- and post-contract services is still alive and healthy. However, at the other end of the spectrum the larger practices are now relabelled as international consulting organisations and would be unrecognisable to the Willises. The principal differences between these organisations and traditional large quantity surveying practices are generally accepted to be the elevation of client focus and business understanding, and the move by quantity surveyors to develop clients' business strategies and deliver added value. As discussed in the following chapters, modern quantity surveying involves working in increasingly specialised and sectorial markets where skills are being developed in areas including strategic advice in the PFI, partnering, value and supply chain management.

From a client's perspective it is not enough to claim that the quantity surveyor and/or the construction industry is delivering a better value service; this has to be demonstrated. Certainly there seems to be a move by the larger contractors away from the traditional low-profit, high-risk, confrontational procurement paths towards deals based on partnering and PFI and the team approach advocated by Latham. Table 1.3 illustrates the trend away from the traditional lump sum contract based on bills of quantities.

The terms of reference for the Construction Industry Task Force concentrated on the need to improve construction efficiency and to establish best practice. The industry was urged to take a lead from other industries, such as car manufacturing, steel making, food retailing and offshore engineering, as examples of market sectors that had embraced the challenges of rising world-class standards and invested in and implemented lean production techniques. *Rethinking Construction* identified five driving forces that needed to be in place to secure improvement in construction and four processes that had to be significantly enhanced, and set seven quantified improvement targets, including annual reductions in construction costs and delivery times of 10 per cent and reductions in building defects of 20 per cent.

The five key drivers that need to be in place to achieve better construction are:

1. Committed leadership.
2. Focus on the customer.
3. Integration of process and team around the project.
4. A quality-driven agenda.
5. Commitment to people.

The four key projected processes needed to achieve change are:

1. Partnering the supply chain – development of long-term relationships based upon continuous improvement with a supply chain.
2. Components and parts – a sustained programme of improvement for the production and the delivery of components.
3. Focus on the end-product – integration and focusing of the construction process on meeting the needs of the end-user.
4. Construction process – the elimination of waste.

The seven annual targets capable of being achieved in improving the performance of construction projects are:

1. To reduce capital costs by 10 per cent.
2. To reduce construction time by 10 per cent.
3. To reduce defects by 20 per cent.
4. To reduce accidents by 20 per cent.
5. To increase the predictability of projected cost and time estimates by 10 per cent.
6. To increase productivity by 10 per cent.
7. To increase turnover and profits by 10 per cent.

The report also drew attention to the lack of firm quantitative information with which to evaluate the success or otherwise of construction projects.

Such information is essential for two purposes:

1. To demonstrate whether completed projects have achieved the planned improvements in performance.
2. To set reliable targets and estimates for future projects based upon past performances.

It has been argued that organisations like the Building Cost Information Service have been providing a benchmarking service for many years through its tender-based index. In addition, what is now required is a transparent mechanism to enable clients to determine for themselves which professional practice, contractor, subcontractor and so on delivers best value.

Although the above figures appear to reinforce the march of design and build there is anecdotal evidence which suggests that as the recession took hold in 2009 clients were reverting to single-stage competitive tendering based on bills of quantities. One of the main reasons cited for this was the poor quality control associated with design and build. Another interesting point about Table 1.3 in particular is the wide fluctuations in some of the statistics, in particular in relation to construction management and partnering.

By 2010 many of the above ingredients of the 1990 heady brew had been factored into UK construction practice.

The rise of the New Engineering Contract (NEC/ECC)

Despite its critics, for many years the default contract recommended by quantity surveyors was the Joint Contracts Tribunal contract, known as the JCT. The main reason for this seems to be that everyone concerned in the construction process is familiar with the JCT, in all its forms, and more or less knows what the outcome will be in the event of a contractual dispute between the parties to the contract. However, the JCT was often blamed for much of the confrontation that has historically been so much a part of everyday life in the construction industry and Latham in his 1994 report recommended the use of the NEC. The NEC was first published in 1993 with a second edition (NEC2), when it was renamed the Engineering and Construction Contract (ECC) and a third edition followed (NEC3) in July 2005. According to the RICS *Contracts in Use Survey* (2007), the NEC was used in 14 per cent of contracts surveyed compared with 61 per cent (by value) that used the JCT. The NEC is now the default contract for many central government agencies, including the Environment Agency and NHS ProCure 21+, and the OGC recommends that public sector procurers use the NEC3 on their construction projects. In the private sector the NEC was used for Heathrow Terminal 5, the Channel Tunnel Rail Link and the Eden Project. One of the main differences between NEC and more traditional forms of contract is that the NEC has deliberately been drafted in non-legal language in the present tense, which may be fine for the parties to the contract but

Table 1.3 Trends in methods procurement – number of contracts

	1985 %	*1987* %	*1989* %	*1991* %	*1993* %	*1995* %	*1998* %	*2001* %	*2004* %	*2007* %
Lump sum – firm BQ	42.8	35.6	39.7	29.0	34.5	39.2	30.8	19.6	31.1	20.0
Lump sum (spec and drawings)	47.1	55.4	49.7	59.2	45.6	43.7	43.9	62.9	42.7	47.2
Design and build	3.6	3.6	5.2	9.1	16.0	11.8	20.7	13.9	13.3	21.9
Construction management	—	—	0.2	0.2	0.4	1.3	0.8	0.5	0.9	1.1
Partnering	—	—	—	—	—	—	—	0.6	2.7	2.4
Others	6.5	5.4	5.2	2.5	3.5	4.0	3.8	2.5	9.3	7.4

Source: RICS Contracts in Use in (2007).

Table 1.4 Trends in methods procurement – value of contracts

	1985 %	*1987* %	*1989* %	*1991* %	*1993* %	*1995* %	*1998* %	*2001* %	*2004* %	*2007* %
Lump sum – firm BQ	59.3	52.1	52.3	48.3	41.6	43.7	28.4	20.3	23.6	13.2
Lump sum (spec and drawings)	10.2	17.7	10.2	7.0	8.3	12.2	10.0	20.2	10.7	18.2
Design and build	8.0	12.2	10.9	14.8	35.7	30.1	41.4	42.7	43.2	32.6
Construction management	—	—	6.9	19.4	3.9	4.2	7.7	9.6	0.9	9.6
Partnering	—	—	—	—	—	—	—	1.7	6.6	15.6
Others	6.5	16.0	29.7	10.5	10.5	9.8	12.5	5.5	15.0	10.8

Source: RICS Contracts in Use (2007).

may cause concern to legal advisers who have to interpret its effect. Another innovation is the inclusion of a risk register which, although it enables the early identification of risks, has led to concerns that it may be skewed in the contractor's favour, obliging the project manager to cooperate to the contractor's advantage. To date it appears as though the NEC is a step in the right direction for construction. There are few disputes involving the NEC that have reached the courts and there is no substantive NEC case law, but time will tell whether that continues to be the situation when it becomes more widely adopted.

New challenges

As the second edition of *New Aspects of Quantity Surveying Practice* went to print in 2005, sustainability and green issues were just coming to prominence in the world as a whole and the construction industry in particular. In June 2007 the RICS published a guide entitled *Surveying Sustainability* which attempted to clarify for the professional the many issues surrounding the topic. As well as this guide a number of other government publications and targets have been and continue to be issued, which, taken together, make addressing sustainability a must for quantity surveyors. Sustainability is so important for the construction industry because construction has been identified as one of the major contributors to carbon emissions and therefore to the great global warming debate, whether or not one actually subscribes to the various theories relating to climate change (Figure 1.4).

Sustainability and green issues will be discussed in more detail later in this book as there is no doubt that it is a topic that will increasingly shape the ways in which buildings are both designed and procured, and therefore the day-to-day life of quantity surveyors when they are consulted about cost advice/implications on green-related matters.

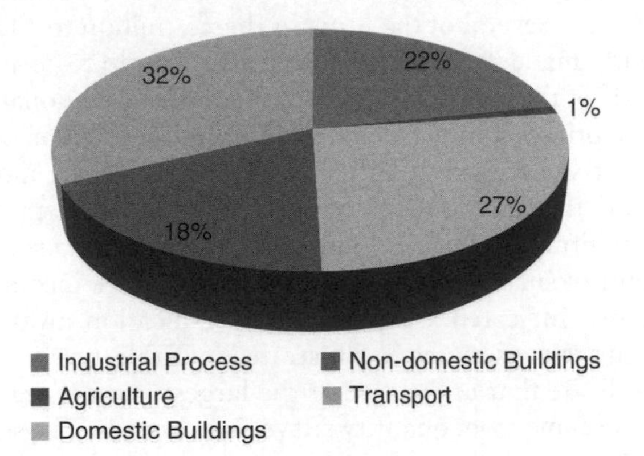

Figure 1.4 Carbon emissions by major sectors

Another matter that has been exercising the mind of quantity surveyors during the recent past is the very future of measurement and the Bill of Quantities. In 2006 the RICS unilaterally announced that 'nobody does measurement any more' and consequently the SMM7 for some elements of the profession had become redundant. The argument went that owing to a lack of recognised industry standard a new approach was required: the New Rules of Measurement (NRM). A more detailed discussion of the NRM is given in Chapter 2.

Conclusion

There can be no doubt that the pressure for change within the UK construction industry and its professions, including quantity surveying, is unstoppable, and that the volume of initiatives in both the public and private sectors to try to engineer change grows daily. The last decade of the twentieth century saw a realignment of the UK's economic base. Traditional manufacturing industries declined while services industries prospered, but throughout this period the construction industry has remained relatively static, with a turnover compared to GDP of 10 per cent. The construction industry is still therefore a substantial and influential sector and a major force in the UK economy. Perhaps more than any other construction profession, quantity surveying has repeatedly demonstrated the ability to reinvent itself and adapt to change.

Is there evidence that quantity surveyors are innovating and developing other fields of expertise? In 2004 a report was published by the RICS Foundation which came to the conclusion that there was evidence of innovation, especially among the larger practices.

The report was based on a survey of twenty-seven consultants from among the largest in the UK, ranked by the number of chartered quantity surveyors employed. The report concluded that there is a clear divide between the largest firms, each generating an income of more than £30 million per annum, and the other firms surveyed. Several of the firms in the £5 million to £15 million fee bands had recently made the transition from partnership to corporate status, while around half of the firms surveyed retained their traditional partnership structure. For the private limited companies this had resulted in organisations of a very different shape, with a flatter structure permitting more devolved responsibility and the potential for better communication throughout the organisation. The firms were asked to identify what percentage of fee income came from 'quantity surveying' services and all other fee income-generating services; the results indicated a significant diversification away from traditional quantity surveying services, as illustrated in Table 1.5.

The results indicate that in the case of the largest firms fewer than 50 per cent of fee income came from quantity surveying services. The services being offered by the firms include project management, legal services, taxation advice, value management and PFI consultancy.

Table 1.5 Results of a study on quantity surveying services

Annual fee income £ million	*% from quantity surveying services Mean*	*Min*	*Max*
>30	49	29	80
20–30	5	5	5
15–20	—	—	—
10–15	63	25	90
5–10	66	36	86
<5	64	34	95

Source: RICS Foundation (2004).

May 2010 saw the formation of the first coalition government in Westminster since the Second World War. What this will mean for construction and the professions is not yet clear, although one thing seems certain: the Private Finance Initiative is to undergo some radical reorganisation and public infrastructure works will have their funding reduced. The remainder of this book will attempt to review the new opportunities that are presenting themselves to the quantity surveyor in a swiftly changing global construction market. It is not the object of this book to proclaim the demise of the traditional quantity surveyor practice offering traditional quantity services – these will continue to be in demand – but rather to outline the opportunities that are now available for quantity surveyors to move into a new era offering a range of services and developing new expertise.

Bibliography

Agile Construction Initiative (1999). *Benchmarking the Government Client*. Stage Two Study. HMSO.

Auditor General (2001). *Modernising Construction*. HMSO.

Banwell, Sir H. (1964). *Report of the Committee on the Placing and Management of Contracts for Building and Civil Engineering Work*. HMSO.

Building Design Partnership (2004). *Learning from French Hospital Design*. Building Design Partnership.

Building/MTI (1999). *QS Strategies 1999, Volumes 1 and 2*. Building/Market Tracking International.

Burnside, K. and Westcott. A. (1999). *Market Trends and Developments in QS Services*. RICS Research Foundation.

CABE (2002). *Improving Standards of Design in the Procurement of Public Buildings*. Commission for Architecture & the Built Environment, London.

Construction Industry Computing Association (2000). *Architectural IT Usage and Training Requirements*. http://www.cica.org.uk.

Cook, C. (1999). QS's in revolt. *Building*, 29 October, p. 24.

CRINE (1994). *Cost Reduction Initiative for a New Era*. United Kingdom Offshore Operators Association.

Department of the Environment, Transport and the Regions (1998). *Rethinking Construction*. HMSO.
Department of the Environment, Transport and the Regions (2000). *KPI Report For The Minister for Construction*. HMSO.
Edkins, A.J. and Winch, G.M. (1999). *Project Performance in Britain and France: The Case of Euroscan*, Barlett Research Paper 7.
Emmerson, Sir H. (1962). *Survey of Problems before the Construction Industries*, HMSO.
Financial News (2000). Report finds majors shunning traditional work. *Building*, 24 November, p. 21.
Goodall, J. (2000). *Is the British Construction Industry Still Suffering from the Napoleonic Wars?* Address to National Construction Creativity Club, London, 7 July.
Hoxley, M. (1998). *Value for Money? The Impact of Competitive Fee Tendering on the Construction Professional Service Quality*. RICS Research.
Latham, Sir M. (1994). *Constructing the Team*. HMSO.
Leibfried, K.H.J and McNair, C.J. (1994). *Benchmarking: A Tool for Continuous Improvement*. Harper Collins.
Lema, N.M. and Price, A.D.F. (1995). Benchmarking performance improvement towards competitive advantage. *Journal of Management of Engineering*, 11, 28–37.
Levene, Sir P. (1995). *Construction Procurement by Government. An Efficiency Scrutiny*. HMSO.
National Audit Office (2001). *Modernising Construction*. HMSO.
Office of National Statistics (2009). *Labour Market Report*. HMSO.
Pullen, L. (2001). What is best value in construction procurement? *Chartered Surveyor Monthly*, February.
Royal Institution of Chartered Surveyors (1971). *The Future Role of the Quantity Surveyor*. RICS.
Royal Institution of Chartered Surveyors (1983). *The Future Role of the Quantity Surveyor*. RICS
Royal Institution of Chartered Surveyors (1991). *Quantity Surveying 2000 – The Future Role of the Chartered Quantity Surveyor*. RICS.
Royal Institution of Chartered Surveyors (1998). *Challenge of Change: QS Think Tank*. RICS.
Simon, Sir E. (1944). *The Placing and Management of Building Contracts*. HMSO.
Thompson, M.L. (1968). *Chartered Surveyors: The Growth of a Profession*. Routledge & Kegan Paul.
Willis, A.J. and Willis, C.J. (1979). *Practice and Procedure for the Quantity Surveyor*. Blackwell.
Winch, G.M. (1996). *The Channel Tunnel Fixed Link*. UMIST Case Study No. 450.

Websites

National Audit Office – http://www.nao.gov.uk
SIMAP – http://simap.eu.int
Tenders Electronic Daily (TED) – http://ted.eur-op.eu.int
Treasury – http://www.hm-treasury.gov.uk

Chapter 2

A new approach to cost advice and measurement

Many within the construction industry regard measurement and the quantity surveyor as synonymous. For years, quantity surveyors, using a variety of standard methods, have provided quantities and schedules for the industry to calculate estimates, tenders and final accounts. However, the popularity of measurement was due in the main to the widespread use of bills of quantities, which during the past few years have declined, while other methods of procurement, not dependent on a detailed bill of quantities, have become more popular with clients. Once a key element of quantity surveying diploma and degree courses, measurement has now been confined to the status of just another module.

The end of the Standard Method of Measurement?

Standard Methods of Measurement of Building Works

For nearly a hundred years the Standard Method of Measurement has been the industry model for preparing bills of quantities. It first appeared in 1922 and was based on 'the practice of the leading London quantity surveyors'. It was an attempt to bring uniformity to the ways by which quantity surveyors measured and priced building works. The seventh edition appeared in 1988 as a joint publication between the RICS and the Building Employers Confederation (BEC) being revised in 1998. The UK Standard Method of Measurement of Building Works (SMM7) has been used as the basis for the preparation of methods of measurement that are used in Malaysia and Hong Kong. The RICS is currently conducting an initiative to produce a new method of measurement for both trade/package measurement as well as estimating and cost planning.

The co-ownership of the Standard Method of Measurement will play a key role in the status and adoption of the New Rules of Measurement since, like a married couple getting divorced, the RICS and the BEC traded insults and press releases. Other methods of measurement that have been developed to cater for particular industry/trade practices are discussed below.

Civil Engineering Standard Method of Measurement (CESMM3)

Sponsored and published by the Institution of Civil Engineers the first edition of CESMM3 appeared in 1976 in order to 'standardise the layout and contents of Bills of Quantities and to provide a systematic structure'. The current, third edition was published in 1991.

The two standard methods of measurement described above reflect the different approaches of the two industries, not only in the nature of the work, but also the degree of detail and the estimating conventions used by both sectors. This in turn reflects the different ways in which building and civil engineering projects are organised and carried out. In general the SMM7 places more emphasis on detail, whereas the CESMM3 takes a more inclusive approach to the measurement process. Building work comprises many different trades, whereas civil engineering works consists of large quantities of a comparatively small range of items. For example, when measuring excavation using SMM7 it is necessary to treat excavation, earthwork support and working space as separate items, whereas when using CESMM3 all these items are included in a single unit.

Other methods of measurement include:

- Standard method of measurement for highways.
- Standard method of measurement for roads and bridges.
- Standard method of measurement for industrial engineering construction, which provides measurement principles for the estimating, tendering, contract management and cost control aspects of industrial engineering construction.
- RICS international method of measurement.

  A useful but sometimes overlooked guide to measurement, now incorporated within the NRM is the RICS Code of Measurement Practice.

The RICS Code of Measurement Practice (sixth edition 2007)

The purpose of the Code is to provide succinct, precise definitions to permit the accurate measurement of buildings and land, the calculation of the sizes (areas and volumes) and the description or specification of land and buildings on a common consistent basis. It is particularly useful when preparing cost plans and giving cost advice. The code is intended for use in the UK only and includes three core definitions that are used in a variety of situations (as noted) as follows.

Gross external area (GEA)

This approach to measurement is recommended for:

- Building cost estimation for calculating building costs for residential property for insurance purposes.
- Town planning applications and approvals.
- Rating and council tax bands.

Gross internal area (GIA/GIFA)

This approach to measurement is recommended for:

- Building cost estimation.
- Marketing and valuation of industrial buildings, warehouses, depart ment stores.
- Valuation of new homes.
- Property management – apportionment of service charges.

GIA is the area of a building measured to the internal face of the perimeter walls at each floor level.

Net internal area (NIA)

This approach to measurement is recommended for:

- Marketing and valuation of shops and supermarkets and offices.
- Rating shops.
- In property management it is used for the apportionment of service charges.

NIA is the usable area within a building measured to the internal face of the perimeter walls at each floor level.

The RICS New Rules of Measurement (NRM)

Background

As previously stated in the introduction to this chapter the Standard Method of Measurement of Buildings Works (SMM), described by RICS Books as a 'landmark publication', currently in its seventh edition, has, for as long as anyone can remember, been the standard set of rules and measurement conventions that quantity surveyors refer to when preparing bills of quantities. Traditionally, this is a document that is drawn up by both sides of the construction industry, from the client side, the RICS and from the contracting side, the Construction Confederation. It is fair to say that the SMM in all its editions has benefited from a captive market, and royalties from the SMM over the years must have been substantial. Imagine the outrage therefore at the Construction Confederation when in June 2006 the RICS announced that it was going alone and would draw up and publish new Rules of Measurement. The reason behind this move was a belief on the part of the RICS that 'Nobody produces bills of quantities and more'. This statement was later clarified as meaning that measurement was taking place, but in a variety of differing formats and conventions. One of the consequences of changing practice was that the Building Cost Information Service was finding it increasingly difficult to obtain cost data from the industry in a form that was meaningful. In 2003 the RICS Quantity Surveying and Construction Professional Group commissioned a report entitled *Measurement Based Procurement of Buildings* from the Building Cost Information Service. The report was based on a survey of practising consultants and contractors, and was carried out during 2002; as with most surveys of this nature the response rate was low, with only 20 per cent of consultants and 12 per cent of contractors responding. The report came to a number of conclusions; it confirmed that measurement still had an important part to play in the procurement of buildings, but perhaps not surprisingly it found that measurement was used by clients, contractors, subcontractors and suppliers in the procurement process in a variety of ways. Following the report, the group established a steering committee with a brief to identify and develop client-focused common standards of measurement at each level of the hierarchy in the construction process, namely:

- Estimating and cost planning
- Procurement
- Maintenance and operation (see Figure 2.1).

The report also concluded that the rise in the use of design-and-build procurement has encouraged the use of contractors' bills of quantities where few documents are prepared to a standard format, for example, in accordance with SMM7. Further, the allegation is that the SMM7 is out of date and represents a time when bills of quantities and tender documents were required to be measured in greater detail than is warranted by current procurement practice and therefore a new approach was required. The RICS *Contracts in Use Survey* (2007) commented on the traditional approach to measurement:

> Bills of quantities refuse to die. It should be noted that while the use of the 'with quantities forms' have declined there are still SMM7 Bills of Quantities being measured. It will be interesting in future years to see what is happening in the market when the RICS publishes NRM2, the procurement section of the NRM suite of documents.

The plan is to publish the New Rules of Measurement in three volumes:

- Volume 1: *Order of Cost Estimating and Elemental Cost Planning.*
- Volume 2: *Construction Quantities and Works Procurement* (expected 2011).
- Volume 3: *Maintenance and Operation Cost Planning and Procurement* (expected 2011/2012).

The status of the Rules of Measurement is the same as other RICS guidance notes and, as such, if an allegation of professional negligence is made against a surveyor, a court is likely to take account of the contents of any relevant guidance notes published by the RICS in deciding whether or not the surveyor had acted with reasonable competence.

The first volume, *Order of Cost Estimating and Elemental Cost Planning*, was launched in March 2009. Some 310 pages in extent, it aims to provide a comprehensive guide to good cost management of construction projects; however, it is Volume 2, dealing with works procurement, or measurement, that has caused a furious reaction from the Construction Confederation, which issued a legal challenge in September 2009 on the basis that it will gain no royalties from the sale of the new rules and that they have been drawn up without its input. The Confederation claimed that the new measurement rules are based on information contained in the previous SMM editions and that there are consequently copyright issues arising from the new publication. The New Rules of Measurement have been largely produced in agreement with the trade and subcontracting bodies; main contractors have not been consulted.

However, the launch by the RICS of Volume 1 may have been timely, since the Construction Confederation announced that it would be winding up following the statement from the Confederation of British Industry (CBI), that it would launch a super-body of twenty-four construction bosses called the Construction Council and would lobby on construction issues. In October 2009 The Construction Confederation ceased trading, with pension debts of £20 million.

The rationale for the introduction of the NRM is that it provides:

- A standard set of measurement rules that are understandable by all those involved in a construction project, including the employer, thereby aiding communication between the project/design team and the employer.
- Direction on how to describe and deal with cost allowances not reflected in measurable building work.
- In addition, it was believed that the SMM7 was UK centric and a more universal approach was now required.

The structure of Volume 1, *Order of Cost Estimating and Elemental Cost Planning*, is as follows:

- *Part 1* places rules of measurement in context with the RIBA Plan of Work and the Office of Government Commerce Gateway Process as well as explaining the definitions and abbreviations used in the rules.
- *Part 2* describes the purpose and content of an order of cost estimate and explains how to prepare an order of cost estimate using three prescribed approaches: floor area, functional unit and elemental method.
- *Part 3* explains the purpose and preparation of elemental cost plans.
- *Part 4* contains tabulated rules of measurement for formal cost plans.

  * The OGC developed the OGC Gateway^{TM} Process as part of the Modernisation Agenda to support the delivery of improved public services, and examines programmes and projects at various key decision points throughout the life cycle of delivery. The process is mandatory in central civil government for procurement, IT enabled and construction programmes and projects. For the purposes of this chapter reference will be made to the RIBA Plan of Work.

  At the time of writing the only section of the new rules to be available is Volume 1; therefore discussion will be confined to this volume. Volume 2 is now anticipated in Autumn 2011.

Joined-up cost advice

The basic premise is that SMM7 is no longer fit for purpose for a number of reasons and that in addition does not adequately address changes in procurement strategies. It presents problems when it comes to cost advice, as it does not reflect the modern-day approach to compiling cost plans and creates difficulties in capturing cost data.

The Building Cost Information Service Standard Form of Cost Analysis (SFCA) was first produced in 1961 when the bill of quantities was king, and was subsequently revised in 1969 and 2008 and has been the industry norm for the past forty years. The SFCA is presented in elemental format but in truth it has not really altered its format since its original launch, whereas the industry has move on and changed. Once the NRM has become widely adopted there is obviously a lot of work to be done by the BCIS in order to convert its current format to the NRM format. Figure 2.1 illustrates a project overview of the NRM from which the various strands of the initiative may be identified.

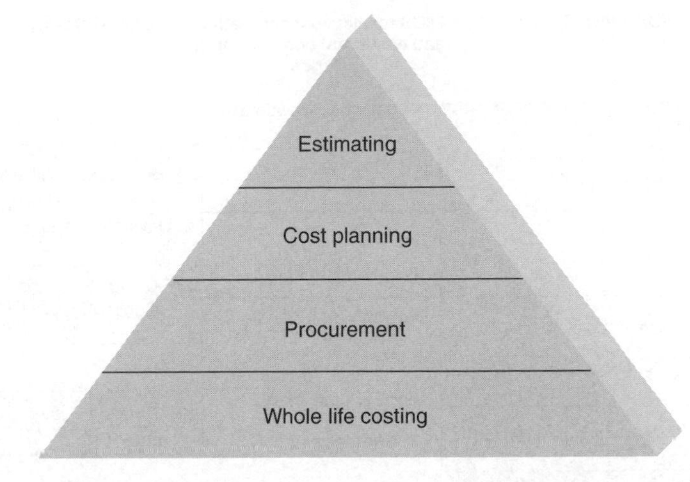

Figure 2.1 A project overview of the NRM

One of the factors that has driven Volume 1 of the NRM is the lack of specific advice on the measurement of building works solely for the purpose of preparing cost estimates and cost plans. As someone who has tried to teach cost planning and estimating for the past forty years I am acutely aware that students, as well as practitioners, are often confused as to how estimates and cost plans should be prepared, resulting in the process taking on the air of a black art! This situation has led to an inconsistent approach, varying from practice to practice, leaving clients a little confused. It is also believed that the lack of importance of measurement has been reflected in the curriculum of degree courses, resulting in graduates being unable to measure or build up rates, a comment not unknown during the past fifty years or so.

As illustrated in Figure 2.2, the process of producing a cost estimate and cost planning has been mapped against the RIBA plan of work and OGC Gateway process. It shows that the preparation and giving of cost advice is a continuous process, that in an ideal world it becomes more detailed as in turn the information flow becomes more detailed. In practice it is probable that the various stages will merge and that such a clear-cut process will be difficult to achieve.

The NRM suggests that the provision of cost advice is an iterative process that follows the information flow from the design team:

- Order of cost estimate
- Formal cost plan 1
- Formal cost plan 2
- Formal cost plan 3
- Pre-tender estimate.

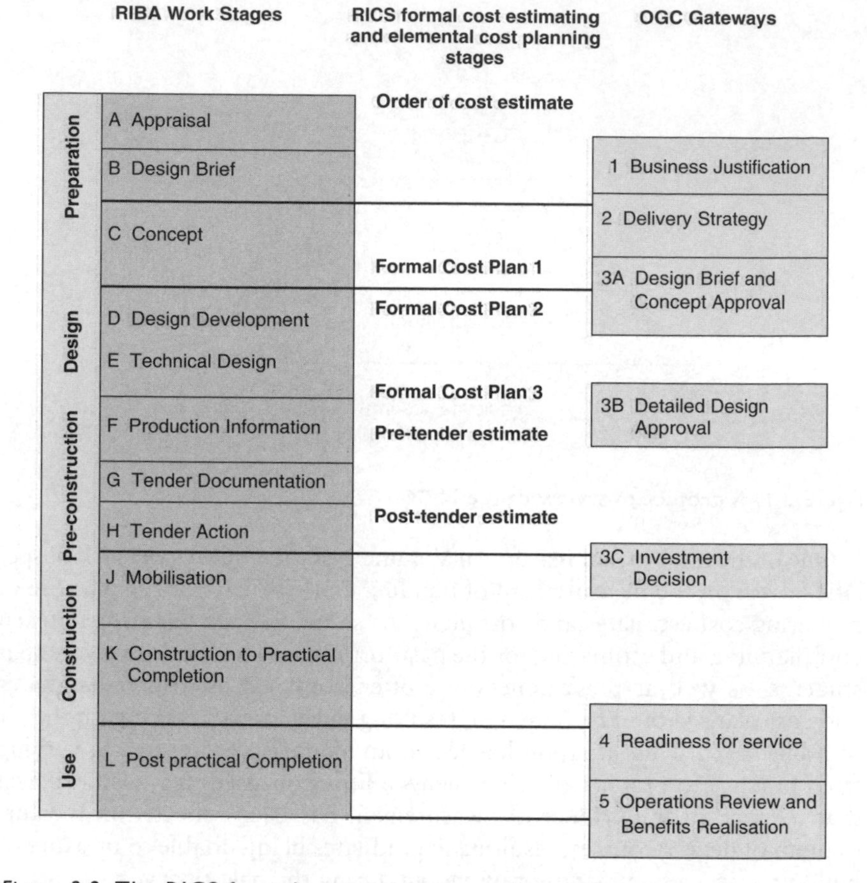

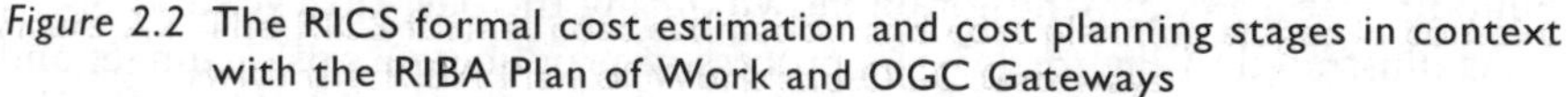

Figure 2.2 The RICS formal cost estimation and cost planning stages in context with the RIBA Plan of Work and OGC Gateways

Source: RIBA Plan of Work is copyright RIBA.

There would therefore appear to be two distinct stages in the preparation of initial and detailed cost advice:

- *Estimate* = an evolving estimate of known factors. Is the project affordable? The accuracy at this stage is dependent on the quality of the information. Lack of detail should attract a qualification on the resulting figures. At this stage information is presented to the client as shown in Table 2.1.
- *Cost plan* = a critical breakdown of the cost limit for the building into cost targets for each element. At this stage it should be possible to give a detailed breakdown of cost allocation as shown in Table 2.2.

In addition, the NRM approach divides cost estimates and cost plans into five principal cost centres:

1. Works cost estimate
2. Project/design team fees estimate
3. Other development/project cost estimate
4. Risk allowance estimate
5. Inflation estimate.

The order of cost information and cost plan stages have differing recommended formats (see Table 2.1 for order of cost estimate recommended format). Compared to the BCIS (SFCA) the NRM format does provide a greater range of cost information to the client, covering the following:

- Building works
- Main contractor's preliminaries
- Main contractor's profit and overheads
- Project/design team fees
- Other development/project costs
- Risk
- Inflation
- Capital allowances, land remediation relief and grants
- VAT assessment.

Table 2.1 RICS New Rules of Measurement – Order of Cost Estimate format

Ref	*Item*	£
1	Building Works	
2	Main Contractor Preliminaries	
3	Main Contractor Overheads and Profit	
	Works Cost Estimate	
4	Project/Design Team Fees	
5	Other Development/Project Costs	
	Base Estimate	
6	Risk Allowances: Design Development Risk Construction Risks Employer Change Risks Employer Other Risks	
	Cost Limit (excluding inflation)	
7	Inflation: Tender Inflation Construction Inflation	
	Cost Limit (including inflation)	

RICS New Rules of Measurement – order of cost estimate format

A feature of the NRM is the detailed lists of information required to be produced by all parties to the process: the employer, the architect, the mechanical and electrical services engineers and the structural engineer must all provide substantial lists of information. There is an admission that the accuracy of an order of cost estimate is dependent on the quality of the information supplied to the quantity surveyor. The more information provided, the more reliable the outcome will be and, in cases where little or no information is provided, the quantity surveyor will need to qualify the order of cost estimate accordingly.

The development of the estimate/cost plan starts with the order of cost estimate.

Works cost estimate stages 1–3

The works cost estimate comprises three constituents:

Works cost estimate → Building works estimate
Works cost estimate → Main contractor's preliminaries
Works cost estimate → Main contractor's overhead and profit

At this stage the main contractor's preliminaries and overheads and profit are included as a percentage, with subcontractors' preliminaries and overheads and profit being included in the unit rates applied to building works.

Perhaps the most worthwhile feature of the NRM is the attempt to establish a uniform approach to measurement based on the RICS Code of Measuring Practice (sixth Edition 2007) in which there are three prescribed approaches for preparing building works estimates:

1. Cost per m^2 of floor area:
   - Gross external area (GEA)
   - Gross internal area (GIA/GIFA)
   - Net internal area (NIA).
2. Functional unit method: building works estimate = number of functional units x cost per functional unit. A list of suggested functional units is included in the New Rules of Measurement, Appendix B.
3. The elemental method: building works estimate = sum of elemental targets. Cost target (for element) = element unit quantity (EUQ) x element unit rate (EUR). Figure 2.2 shows the amount of detail required for this approach, although the choice and the number of elements used to

break down the cost of building works will be dependent on the information available. Rules for calculating EUQ are included in Appendix E of the NRM.

At this stage the main contractor's preliminaries and profit and overheads are recommended to be included as a percentage addition. Subcontractors' overheads and profit should be included in the unit rates applied to building works.

Project and design fees stage 4

In the spirit of transparency the costs associated with project and design fees are also itemised:

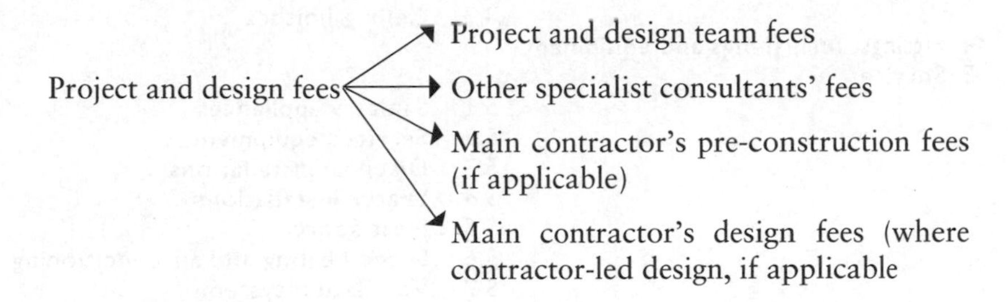

Other development and project cost estimates stage 5

This section is for the inclusion of costs that are not directly associated with the cost of the building works, but form part of the total cost of the building project, for example, planning fees.

Risk allowance estimate stage 6

Risk is defined as *the amount added to the base cost estimate for items that cannot be precisely predicted to arrive at the cost limit.*

The inclusion of a risk allowance in an estimate is nothing new; what perhaps is new however is the transparency with which it is dealt with in the NRM. It is hoped therefore that the generic cover-all term 'Contingencies' will be phased out. Clients have traditionally homed into contingency allowances, wanting to know what the sum is for and how it has been calculated. The rate allowance is not a standard percentage and will vary according to the perceived risk of the project. Just how happy quantity surveyors will be to be so up front about how much has been included for unforeseen circumstances or risk will have to been seen. It has always been believed by many in the profession that carefully concealing pockets of money within an estimate for extras or additional expenditure is a core skill.

Table 2.2 RICS New Rules of Measurement formal cost plan format

Group element	*Element*
1 Substructure	
2 Superstructure	
	2.1 Frame
	2.2 Upper floors
	2.3 Roof
	2.4 Stairs and ramps
	2.5 External walls
	2.6 Windows and external doors
	2.7 Internal walls and partitions
	2.8 Internal doors
3 Internal finishes	
	3.1 Wall finishes
	3.2 Floor finishes
	3.3 Ceiling finishes
4 Fittings, furnishings and equipment	
5 Services	
	5.1 Sanitary appliances
	5.2 Services equipment
	5.3 Disposal installations
	5.4 Water installations
	5.5 Heat source
	5.6 Space heating and air conditioning
	5.7 Ventilation systems
	5.8 Electrical installations
	5.9 Gas and other fuel installations
	5.10 Lift and conveyor installations
	5.11 Fire and lightning protection
	5.12 Communication, security and control systems
	5.13 Special installations
	5.14 Builders' work in connection with services
	5.15 Testing and commissioning of services
6 Complete buildings and building units	
7 Works to existing buildings	
8 External works	
9 Facilitating works	

So how should risk be assessed at the early stages in the project? It is possible that a formal risk assessment should take place, and this would be a good thing, using some sort of risk register. Obviously, the impact of risk should be revisited on a regular basis as the detail becomes more apparent.

Risks are required to be included under four headings:

- Design development risks; for example, design development and environmental issues.
- Construction risks; for example, site restrictions, existing services.
- Employer's change risk; for example, changes in the scope of the works or brief.
- Employer's other risks; for example, early handover, postponement/acceleration.

Inflation estimate stage 7

Finally, an allowance is included for inflation under two headings:

- Tender inflation: an allowance for the period from the estimate base date to the return of the tender.
- Construction inflation: to cover increases from the date of the return of tender to a mid-point in the construction process.

Inflation should be expressed as a percentage using either the retail price index, tender price index or the BCIS building cost indices. This adjustment is of course in addition to any price adjustments made earlier in the process when adapting historic cost analysis data. In addition, care should be taken not to update previous rates that were based on percentage additions (e.g. main contractor's preliminaries, main contractor's overheads and profit and project/design team fees), as these will be adjusted automatically when the percentages are applied.

Finally, it is suggested that other advice could be included relating to:

- value-added tax
- capital allowances
- land reclamation relief
- grants.

Whether this will be possible remains to be seen; certainly giving tax advice has been a stock in trade for many quantity surveyors for some time, but it really needs specialist, up-to-date information. In addition, particularly with VAT, the tax position of the parties involved may differ greatly and advice should not be given lightly.

From this point on advice is given by the preparation of formal cost plans 1, 2 and 3. It is anticipated that for the formal cost plan stages the elemental approach should be used and this should be possible, since the quantity and quality of information available to the quantity surveyor should be constantly increasing. Table 2.3 demonstrates the degree to which detail increases during this process.

Table 2.3 The preparation of formal cost plans

LEVEL 1	*LEVEL 2*	*LEVEL 3*
Group element	*Element*	*Sub-element*
3. Internal finishes	1. Wall finishes	1. Finishes to walls
	2. Floor finishes	1. Finishes to walls 2. Raised access floors
	3. Ceiling finishes	1. Finishes to ceilings 2. False ceilings 3. Demountable special ceilings

At the formal cost plan stages the NRM recommends that cost advice is given on an elemental format and to this end Part 4 of the NRM contains comprehensive rules for the measurement for building works; tabulated rules of measurement for elemental cost planning, enable quantities to be measured to the nearest whole unit, provided that this available information is sufficiently detailed. When this is not possible, measurement should be based upon GIFA. From formal cost plan stage 2 cost checks must be carried out against each pre-established cost target based upon cost-significant elements. One thing that is clear is that the NRM approach, if followed, appears to be fairly labour intensive, and one problem is that the cost planning stages and procurement document stages will morph so that the final cost plan becomes the basis of obtaining bids. Over the coming years it will be interesting to learn to what extent the NRM replaces the tried-and-trusted standard methods of measurement not only in the UK but in the overseas market.

Bibliography

Badke, E. (2008). Are you measuring up?, *RICS Construction Journal*, June/July, pp. 15–17.

Martin, J. (2003). *e-Procurement and Extranets in the UK Construction Industry*. BCIS.

RICS (2009). RICS New Rules of Measurement, Volume 1, *Order of Cost Estimating and Elemental Cost Planning, 1st Edition*. RICS Books.

Chapter 3

Sustainability and procurement

> Sustainable procurement – in short using procurement to support wider social, economic and environmental objectives, in ways that offer real long term benefits.
>
> Sir Neville Simms, Chairman,
> UK Sustainable Procurement Taskforce 2006

Where has sustainability come from?

During the preparation of previous editions of this book, sustainability and green issues were discussed by few people in the construction industry in the course of day-to-day business. Sustainability, rather like stress, appears to have crept up on the United Kingdom construction industry over the past twenty years or so. Yet concerns about climate change and the environment can be traced back several centuries, although it was not until the late 1960s when organisations such as Greenpeace were formed that it attracted public attention. There followed a number of reports and protocols, the most notable being:

- The Brundtland Report (1987), also known as *Our Common Future*, which linked sustainability and development, and established the concept of triple bottom-line sustainability; environment, social and economic forces.
- The Rio Declaration (1992) which established the concept of *the polluter pays*.
- The Kyoto Protocol (1997) agreed under the United Nations Framework Convention on Climate Change.

Opinion about environmental impact gathered momentum in the decade after Kyoto with several severe warnings, including the Stern Report (2005), indicating that human activity was the primary cause of climate change and that urgent action was required to change behaviour. Stern's report, together with high-profile media coverage of the former American Vice President Al

Gore's endorsement of the film *An Inconvenient Truth*, convinced corporate and government opinion of the necessity to take green issues seriously. A recent report by the National Audit Office, *Improving Public Services Through Better Construction*, concluded that £2.6 billion per annum is still wasted through a variety of reasons including lack of consideration of whole-life cost, sustainability and green issues.

One of the major obstacles to the introduction of more sustainable design and construction solutions is the perception that to do so will involve additional costs – typically 10 per cent on capital costs. However, a report by BRE Trust and Cyril Sweett entitled *Putting a Price on Sustainability* (2005) appears to demonstrate that this need not necessarily be the case. In fact the report points out that significant improvements in the sustainability performance of buildings need not be expensive, although it is still true that in order to attain high EcoHomepoint or BREEAM ratings (see below for definitions) there is the need incur significant up-front investment. Nevertheless, the general uncertainty over the cost impact to an entire development's profitability could deter cost-adverse funders from backing a green project.

Various attempts have been made to define the term 'sustainable or green construction'. In reality it would appear to mean different things to different people in different parts of the world, depending on local circumstances. Consequently, there may never be a consensus view on its exact meaning. However, one way of looking at sustainability is 'The ways in which built assets are procured and erected, used and operated, maintained and repaired, modernized and rehabilitated and reused or demolished and recycled constitutes the complete life cycle of sustainable construction activities.' In 2005 the RICS announced that it was establishing a new commission with a mission to 'Ensure that sustainability becomes and remains a priority issue throughout the profession and RICS'. In general a sustainable building reduces the impact upon the environmental and social systems that surround it as compared to conventional buildings; that is to say, green buildings use less water and energy, as well as fewer raw materials and other resources.

Why should sustainability concern the quantity surveyor? Here are some statistics:

- Buildings are the single most important contribution to greenhouse gas emissions with the construction sector being responsible for one-sixth of the total freshwater withdrawals and, taking into account demolition, generates 30 per cent of waste in OECD countries. In addition, around 40 per cent of total energy consumption and greenhouse gas emissions are directly attributable to constructing and operating buildings according to Energy Action. Measured by weight, construction and demolition activities also produce Europe's largest waste stream – between 40 per cent and 50 per cent – most of which is recyclable.

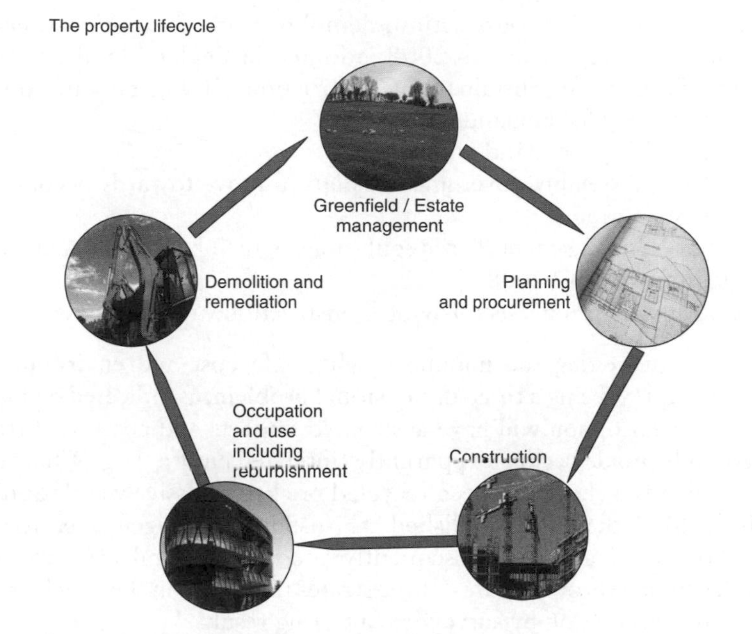

Figure 3.1 The property life cycle.

- Contractors, particularly those involved with public/private partnerships, are beginning to recognise the importance of sustainability issues and the early consideration of whole-life costs.
- Construction clients are increasingly realising the marketing potential of green issues; for example:
  - increase in rental prices
  - increase in occupancy rates
  - reduction in running costs
  - improved productivity
  - 30 per cent of newly built or renovated buildings suffer from sick building syndrome.

There is little wonder that the construction industry and its associated materials and manufacturing sectors has been singled out for action in the green debate when the statistics are laid out!

An estimated 25 million tonnes of construction waste end up in landfill without any form of recovery or reuse. UK governments have set out to reduce construction waste to landfill for economic and environmental reasons through a number of initiatives, including:

- A target for halving construction, demolition and excavation waste to landfill by 2012, relative to 2008, adopted in England by the government's Strategy for Sustainable Construction (2008), building on the Waste Strategy for England (2007).
- The Zero Waste Scotland policy goal.
- The Welsh Assembly government's plan to move towards becoming a zero waste nation.
- Site Waste Management Plan Regulations, which became mandatory in England from April 2008.
- The Strategic Forum's sector-wide Construction Commitments.

The process of getting the minimum whole-life cost and environmental impact is complex, being a three-dimensional problem, as indicated by Figure 3.2. Each design option will have associated impacts and costs, and trade-offs have to be made between apparently unrelated entities (e.g. What if the budget demands a choice between recycled bricks or passive ventilation?).

In June 2007 the RICS published its sustainability agenda entitled *A Green Profession?* and has subsequently produced several practice notes and guides for surveyors. Figure 3.1 illustrates the points in the development cycle where intervention by surveyors can bring results.

There would appear to be a strong business case for the sustainable construction agenda, based upon:

- Increasing profitability by using resources more efficiently.
- Firms securing opportunities offered by sustainable products or ways of working.
- Enhancing company image and profile in the marketplace by addressing issues relating to corporate and social responsibility.

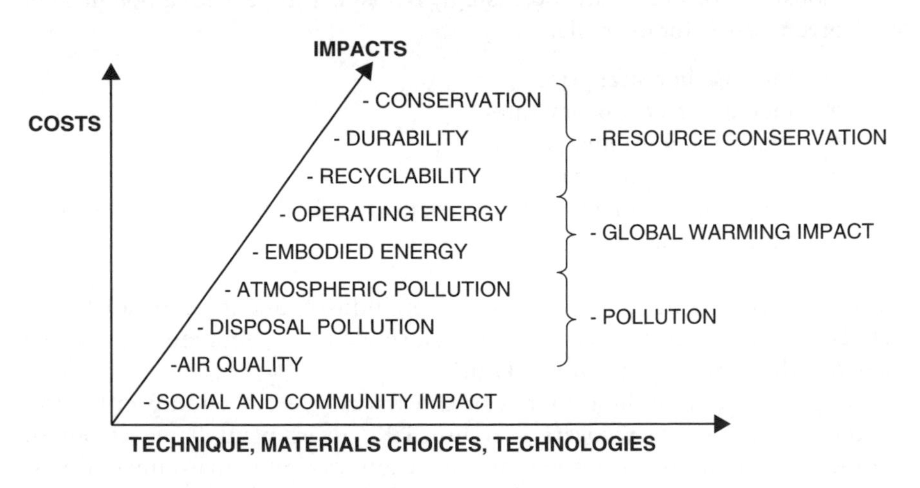

Figure 3.2 Techniques, material choices and technologies

However, despite the above, sustainability/green issues remain a nebulous area for many within the construction industry. Popular perception is that there is a lack of customer demand for sustainability to be considered during the design and procurement stages; however, consider the following reasons for greening:

- English partnerships require developers to achieve a minimum BREEAM or EcoHomes star rating.
- Public sector contractors must achieve BREEAM excellence for all new buildings.
- Many high-profile private developers and landowners are seeking the same or higher standards of sustainability performance from their partners.
- Investors are becoming increasingly interested in sustainability and are encouraging property industry partners to do the same.

The implementation of the Energy Performance of Buildings Directive is discussed below.

Some of the benefits of having a BREEAM rating are claimed to be:

- Demonstrating compliance with environmental requirements.
- Marketing: as a selling point to tenants and customers.
- Financial: to achieve higher and increased building efficiency.

Increasingly, clients as well as end-users are requesting improved sustainability performance from their buildings, over and above the regulatory requirements arising from changes in the Building Regulations. Methodologies such as BREEAM and LEED (Leadership in Energy and Environmental Design) are often used as the vehicle for achieving these improvements. However, these tools are largely environmentally biased, and it is important that the wider social and economic dimension of sustainability is also considered- (see Chapter 8). It is strongly recommended that these issues are considered holistically at an early stage in project inception and taken forward in an integrated manner. From a sustainability perspective, refurbishment projects are increasingly expected to achieve design standards expected of new build projects, including:

- improved quality and value for money;
- reduced environmental impact and improved sustainability;
- healthy, comfortable and safe internal and external environments that offer high occupant satisfaction and productivity;
- low costs in use;
- a flexible and future-proofed design.

The introduction of Energy Performance Certificates and Display Energy Certificates as required by the European Union Directive; 2002/91/EC on the Energy Performance of Buildings Directive (EPBD) offers an opportunity to improve the energy performance of the existing building stock, and embarking upon a refurbishment or refit without ensuring that an improvement of energy performance is specified would be ill-advised, as there are significant benefits to be realised both in cost reductions as well as reductions in carbon dioxide emissions. Ultimately property professionals have a role and responsibility to ensure that we are maintaining and indeed contributing to increasing the value of client assets and, in what we believe to be an increasingly resource-constrained and climatically challenging future, building sustainably is a pathway to assuring asset value.

The measures adopted to assess sustainability performance, which developers and design teams are encouraged to consider at the earliest possible opportunity, are:

- BREEAM (Building Research Establishment Assessment Method)
  BREEAM-in-use for existing buildings
  EcoHomes points.

BREEAM

BREEAM has been developed to assess the environmental performance of both new and existing buildings. BREEAM assesses the performance of buildings in the following areas:

- management: overall management policy, commissioning and procedural issues;
- energy use;
- health and well-being;
- pollution;
- transport;
- land use;
- ecology;
- materials;
- water, consumption and efficiency.

In addition, unlike EcoHome points, BREEAM covers a range of building types, for example:

- offices;
- industrial units;
- retail units;
- schools;

- other building types such as leisure centres can be assessed on an ad hoc basis.

In the case of an office development the assessment would take place at the following stages:

- design and procurement
- management and operation
- post-construction reviews
- building performance assessments.

A BREEAM rating assessment comes at a price and according the BRE website the fee scale for BREEAM assessors to carry out an assessment at each of the above stages could be several thousands of pounds per stage, an item that should be considered when completing Section 5 of the RICS New Rules of Measurement: *Order of Estimating and Cost Planning*.

Perhaps one of the most informative pieces of research into the cost of complying with BREEAM was carried out jointly between Cyril Sweett and the BRE. The research tried to dispel the widely held view that improving BREEAM ratings is necessarily an expensive exercise and demonstrated that an increase of between one and three BREEAM rating levels can be achieved at an additional cost of up to 2 per cent of capital cost. The study includes four building types: a house, a naturally ventilated office, an air-conditioned office and a PFI health centre. In order to establish a level playing field some credits were not accessed as they relate to location of the site. This was particularly the case with the PFI health centre. Interestingly, on a baseline cost comparison, only with the PFI project, where the consortia will have to meet the running costs of the building over the concession period, typically thirty years, was there a high percentage of good BREEAM ratings in various locations, with minimal increases in expenditure needed to increase levels to very good or excellent.

A number of key issues were highlighted in the Cyril Sweett publication:

- *Timing*: many BREEAM credits are affected by basic building form and servicing solutions. Cost-effective BREEAM compliance can only be achieved if careful and early consideration is given to BREEAM-related design and specification details. Clear communication between the client, design team members and, in particular, the project cost consultants is essential.
- *Location*: building location and site conditions have a major impact upon the costs associated with achieving very good and excellent compliance.
- *Procurement route*: the PFI and similar procurement strategies that promote long-term interest in building operations for the developer/contractor typically have a position influence on the building's

environmental performance and any costs associated with achieving higher BREEAM ratings.

The Building Regulations (2007) introduced a tougher energy and environmental section, and these new regulations were mandatory from October 2009. In addition, the Climate Change Bill will result in Scotland having the most ambitious climate change legislation anywhere in the world with a mandatory target of cutting emissions by 80 per cent by 2050. Consequently new energy performance standards for buildings and large existing buildings are required and quantity surveyors must be able to provide cost advice on alternative solutions. So how does BREEAM work? BREEAM measures the environmental performance of buildings by awarding credits for achieving a range of environmental standards and levels of performance. Each credit is weighted according to its importance and the resulting points are added up to give a total BREEAM score and rating (Table 3.1).

Location and site conditions have a major impact on the assessment and of course these factors may be outside of the design team's influence.

BREEAM is assessed over several categories (see Figure 3.3). Each category contributes a percentage towards the overall rating

The higher the BREEAM rating the more mandatory requirements there are and the progressively harder they become. In 2008 new BS ISO 15686 – 5 Service Life Planning – Buildings and Constructed Assets Standards were introduced.

EcoHomes points

EcoHomes points are now phased out in England and Wales for new buildings, but they are still used in Scotland as well as the Low Carbon Building Strategy for Scotland (2007). They assess the green performance of houses over a number of criteria:

Table 3.1 BREEAM categories

BREEAM categories	*Current weightings (2008)*
• Management	12%
• Energy	19%
• Water	6%
• Land use and ecology	10%
• Health and well-being	15%
• Transport	8%
• Materials	12.5%
• Waste	7.5%
• Pollution	10%

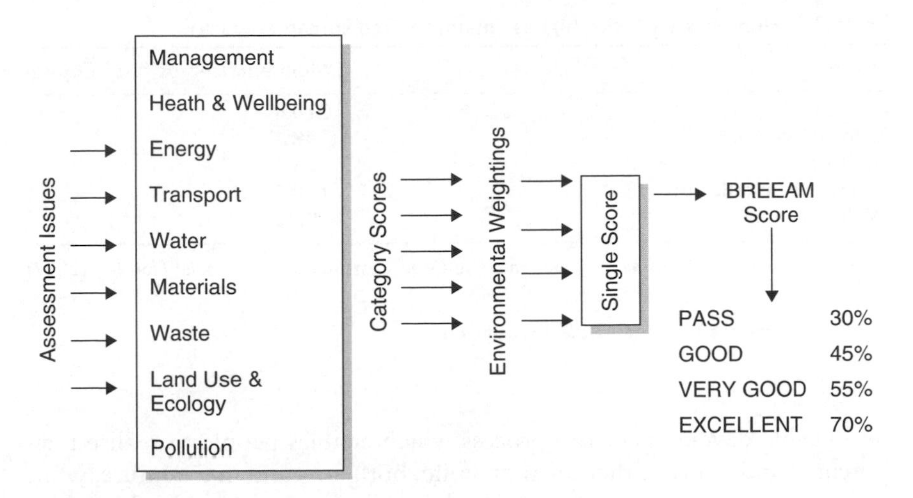

Figure 3.3 BREEAM scoring

- Reducing CO_2 emissions from transport and operational energy.
- Reducing main water consumption.
- Reducing the impact of materials use.
- Reducing pollutants harmful to the atmosphere.
- Improving the indoor environment.

Measurement and costing sustainability

As sustainability is a comparatively new topic for most quantity surveyors, measuring and costing green factors can present a challenge.

The elements with the highest maintenance-to-capital-cost ratios are generally those that are included in major refurbishments, mainly the services elements and the fittings. Conversely external works also has a high ratio because of the constant repletion of fairly small items of work, particularly grass and planted areas, which need constant maintenance at certain times of the year. Table 3.2 shows the elements where maintenance costs as a percentage of capital costs are highest, based on a three-bed, five-person semi-detached house.

These parameters will of course change with different building types and clients.

Sustainable procurement

To understand what sustainable procurement means it is important to first understand what is meant by 'sustainable development' and 'procurement'.

Table 3.2 Elements with the highest maintenance to capital ratio

Element	*Maintenance* as % of Capital*
Fittings	210%
External Works	148%
Heating	133%
Electrical Installation	118%
Wall Finishes	92%

Source: *Guide to the Housing Corporation's Life Cycle Cost Measure for Social Housing* (2007) BCIS.

Note: * NPV over 60 years at 3.5% discount rate.

Sustainable development is a process which enables people to realise their potential and improve their quality of life, both now and in the future, while protecting the environment. Sustainable development policy should include long-term planning, consideration of impacts beyond the local area (regional, national and international impacts) and the integration of social, economic and environmental issues. Procurement is the whole process of acquisition from third parties covering goods, services and capital projects. The process spans the whole life cycle from initial concept through to the end of the useful life of the asset (including disposal) or end of the services contract.

Sustainable procurement is a key method for delivering an organisation's sustainable development priorities. It is all about taking social and environmental factors into consideration alongside financial factors in making these decisions. It involves looking beyond the traditional economic parameters and making decisions based on the whole life-cycle cost, the associated risks, measures of success and implications for society and the environment. Making decisions in this way requires setting procurement in the broader strategic context, including value for money, performance management, and corporate and community priorities.

At the design stage the quantity surveyor needs to be aware of the drivers for sustainability and the impact these have upon capital and life cycle costs, as well as the technical requirements of sustainable buildings, so that these are developed into realistic costs and not arbitrary percentage additions. When the surveyor is required at this stage to liaise with the client and professional team to determine the client's initial requirements and to develop the client's brief, consideration should be given to the client's overall business objectives, particularly any corporate responsibility targets likely to affect the project. In advising the client on demolition and enabling works, the surveyor is advised to consider carrying out a pre-demolition audit to maximise material reclamation and reuse and to minimise waste to landfill. The procurement of demolition and enabling works could include evaluation criteria that consider a company's sustainability credentials. Specialists would be required to contribute to meeting the client's objectives and the project targets in the key sustainability areas.

Where the activities relate to CDM Regulations surveyors need to be aware of site waste management plans. Where the client's objectives include achieving ratings/levels under BREEAM, LEED or the *Code for Sustainable Homes*, surveyors would be expected to familiarise themselves with the specialists who need to be appointed both to carry out the assessment and to provide the necessary reports required by the schemes.

In advising on the cost of the project, sustainability implications of alternative design and construction options need to be understood. It is recommended that cost estimates include cost/m^2 information for indicative low and zero carbon and renewable energy schemes and material selection as required by the RICS new rules of measurement. Costing of issues not generally associated with building design is extremely important (e.g. those actions identified in an Environmental Impact Assessment or the implications of a green travel plan), and the quantity surveyor would be expected to understand or be able to undertake life cycle assessment for the whole development and not just the building. A site visit can identify issues likely to affect cost, time or method of application, including existing buildings on site (including cultural heritage), existing ecological features on site that may need protecting to achieve BREEAM credits, local road layouts that could create traffic congestion and noise, existing watercourses and the implications for storm water control and attenuation, and areas of the site that are liable to flood. Advising on the likely effect of market conditions can involve looking at the possible level of employment and skills in the area, and the levels of crime that might affect the site. The project costs at this stage can influence a financial appraisal, and surveyors are advised to ensure that they understand what is to be priced in order to provide a level of accuracy and avoid substantial cost increases at a later date. In addition to considering the effects of site usage, shape of the building, alternative forms of design, procurement and construction and so on, the surveyor would be expected to be able to proactively advise on the sustainability implications of various low and zero carbon technologies, renewable energy installations and material selections. The surveyor would also be expected to be able to advise on the cost implications of other sustainability issues, including possible construction waste, levels of local employment and skills, and traffic and transport.

When advising on tendering and contractual procurement options, it is recommended that consideration is given in pre-qualification documentation to evaluation of the bidder's response to sustainability issues, particularly those affecting the project. It is important to ensure that the client's and the project's sustainability requirements that were incorporated into the project brief have been reflected in the tender documents, and to ensure that the documentation also includes a responsible approach to sustainability in the contractor's operations, preliminaries and temporary work.

Where bills of quantities are required, it is important that every effort is made to adequately measure all sustainability-related products and technologies, avoiding where possible provisional sums. It is important to ensure that the tender report identifies the sustainability issues/risks affecting the project and the bidder's response to them. Quantity surveyors are advised to carry out an analysis of the contractor's sustainability costs, to compare it with benchmarks and to prepare a report. It is important to ensure that variations in sustainability implications are valued and agreed.

Key to achieving best whole-life value are the following factors:

- understanding value (see Chapter 6)
- assessing value
- putting a cost on value propositions
- identifying the best value-sustainable solution *prior to commitment to invest*
- optimising value over the whole asset life cycle.

From a quantity surveyor's perspective it must be possible to quantify and cost the impact of introducing sustainable practice into construction. In the course of the preparation of the RICS report *A Green Profession?* published in 2007, members in several countries were questioned concerning their attitude to sustainability. Overall responses appeared to show over 40 per cent of those questioned considered green issues to be highly relevant; however, further analysis shows that, not unsurprisingly, the greatest concern is in land professional groups whereas civil engineers, investment managers in the private sector and surveying practices in the private sector were not so concerned. For all respondents the four most important sustainability issues were:

- energy supply (59% of responses)
- land contamination (43%)
- transport (36%)
- waste management (34%).

It was noted by one UK based respondent from the commercial faculty that a building's environmental sustainability is increasingly important to investors. Therefore its energy efficiency, level of carbon use and contribution to global warming are increasingly important, whereas habitat/biodiversity loss, social exclusion and water supply issues are considered to be less important overall.

The two most important barriers which stand out as being of almost equal importance were a lack of knowledge and a lack of expertise (72 per cent and 71 per cent of respondents); this in turn reflects a lack of training and education in relevant techniques. By contrast, the least important barriers to use are an inability to recognise their value (some 29 per cent agreed or strongly agreed), the exclusion of tools from the organisation's policy/strategy (34 per cent agreed or strongly agreed) and being unconvinced of their value (some 35 per cent agreed or strongly agreed).

However, the barriers to green development are at present substantial and include:

- lack of a clear project goals (i.e. targets)
- lack of experience
- lack of commitment
- complicated rating process.

How can attitudes in construction and the profession and expertise in sustainable issues be enhanced? Figure 3.4 illustrates some of the initiatives that need to be put in place in order to increase competencies in sustainable issues.

Of course the majority of building stock is existing, and therefore one of the biggest challenges for the construction industry is how to deal with this stock. For example, approximately one-third of total CO_2 emissions is from commercial buildings of which commercial offices are a major contributor. The actions of both landlords and tenants contribute to building energy performance and in turn the CO_2 emissions produced. The landlord has sole control over the building fabric performance, whereas the tenant is responsible for using the building and controls hours of use, IT equipment and management of the setting of temperatures. In 2009 as part of the IPF Research Programme a group led by Cyril Sweett investigated the cost of making energy-efficient improvements to existing commercial buildings held in investment portfolios. Commercial buildings are of course not homogeneous, having very different characteristics that influence energy consumption and CO_2 emissions. For the IPF study a range of existing office buildings

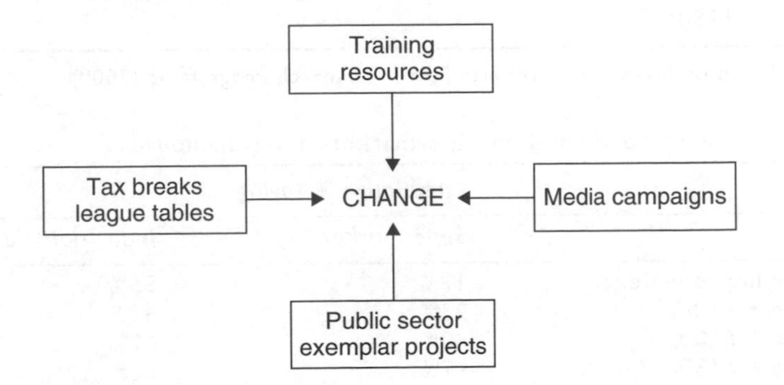

Figure 3.4 Bringing about change

were selected, in the main constructed in the 1990s with a mixture of narrow and deep plans, with and without air conditioning. The research found that in the case of a 1990 built deep plan air-conditioned office building, for a baseline cost of £1000/m^2 reductions of 25 per cent could be achieved by modernising older buildings constructed in accordance with the 1990 Part L Building Regulations. However, by increasing the budget to £1050/m^2, considerably greater reductions could be achieved (see Table 3.3).

As illustrated in Table 3.3, it is estimated that emissions can be reduced by up to 54 per cent by increasing the budget by £150/m^2 from the baseline allowance. It is also estimated that this expenditure would be sufficient to achieve an EPC rating of B for buildings that are starting from a baseline rating of D or E. Similar studies were conducted on two other categories of commercial buildings: supermarkets and industrial warehouses (see Table 3.4).

In all the building types considered the key improvements included:

- Variable speed heating pumps — pay-back period twelve months
- Energy-efficient lighting — pay-back period three to four years
- Air-conditioning fan coil units — pay-back period three to four years
- Condensing boilers — pay-back period eleven years.

Items considered but rejected due to the long pay-back period included wind turbines and high-efficiency chiller units. Retrofit external wall components were also considered and, although this approach improved the comfort and

Table 3.3 Achieving reductions in emissions

	Cumulative % saving
	1990 Deep plan air conditioned office
Base line improvement – £1000/m^2	25%
Budget – £1025/m^2	37%
Budget – £1050/m^2	47%
Budget – £1075/m^2	49%
Budget – £1150/m^2	54%

Source: Based on Investment Property Forum Research Programme (2009).

Table 3.4 Studies conducted on supermarkets and warehouses

	Cumulative % saving	
	Supermarket	*Industrial/Warehouse*
Baseline improvement	12%	35%
Baseline + £10/m^2	23%	41%
Baseline + £100/m^2	40%	62%
Baseline + £130/m^2	47%	66%

Source: Based on Investment Property Forum Research Programme (2009).

environment for the occupants the costs were considerable; this could be accompanied by a potential loss of usable floor space.

In general terms the quantity surveyor should consider the implications of ethical procurement, which encompasses:

1. Whole-life costs/life cycle costs
2. Costs of alternative (renewable) materials
3. Renewable energy schemes
4. Recycled contents schemes
5. The ethical sourcing of materials and labour.

Whole life costs/life cycle costs

The service life of an element, product or whole building may be viewed in one or more of the following ways:

- Technical life – based upon physical durability and reliability properties.
- Economic life – based upon physical durability and reliability properties.
- Obsolescence – based upon factors other than time or use patterns (e.g. fashion).

Examples of common terms used to describe the consideration of all the costs associated with a built asset throughout its lifespan are:

- costs in use
- life cycle costs
- whole-life costs
- through-life costs.

There have been various attempts to provide practising surveyors with information for WLC calculations. There are two major areas:

- The parameters of the study.
- The accuracy of the cost data.

Definitions in the BS ISO 15685:5 are:

- *Life cycle costing*: a methodology for the systematic economic evaluation of the life cycle costs over the period of analysis, as defined in the agreed scope.
- *Whole-life costing*: a methodology for the systematic economic consideration of all the whole-life costs and benefits over the period of analysis, as defined in the agreed scope.

The International Standards Organisation (ISO) defines life cycle costs as 'the cost of an asset throughout its lifecycle while fulfilling the performance requirements'.

Life cycle costing is basically a simple concept – it answers the question: 'If I build this building, what future costs will I be letting myself in for?' Thus it is only a projection of the costs that result from commissioning a building, and which will be the responsibility of the client. While life cycle costs are not difficult, it is complex because potentially there are a huge number of costs to consider. It is also complicated by the introduction of time into the equation and therefore the ways of how to treat the effects of inflation, and lost investment opportunities or money.

Part 5 covers life cycle costing:

- Definitions, terminology and abbreviations.
- Principles of life cycle costing (i.e. purpose and scope; what costs to include/exclude; typical forms and level of LCC analysis at key stages; outputs).
- Forms of LCC calculations and six methods of economic evaluation (with informative examples).
- Setting the scope for LCC studies including how to deal with risks and uncertainty.
- How LCC forms part of the whole-life costing business investment option appraisal process.
- Reporting and analysis techniques.

The ISO will allow organisations to build up life cycle costs of construction projects on a common basis internationally. At present, there is no way to compare LCC estimates, and few organisations are able to estimate LCC. The ISO will eliminate confusion in the industry and is likely to become the established methodology going forward.

First introduced to the UK construction industry over four decades ago by Dr P.A. Stone as costs in use; it is only recently, with the widespread adoption of public/private partnerships in the early 2000s as the preferred method of procurement by the majority of public sector agencies, that the construction industry has started to see some merit in whole-life costing. In addition, building owners with long-term interests in property are starting to demand evidence of the future costs of ownership. For example, PFI prison projects are commonly awarded to a consortium on the basis of Design, Build, Finance and Operate (DBFO), and contain the provision that, at the end of the concession period, typically twenty-five years, the facility is handed back to HM Prisons in a well-maintained and serviceable condition. This is of course in addition to the operational and maintenance costs that will have been borne by the consortia over the contract period. Therefore, for PFI consortia, given the obligations touched upon above, it is clearly in the consortia's interest to give rigorous attention to costs incurred during the proposed assets life cycle in order to minimise operational risk. Although Stone's work was well received in academic circles, where today extensive research still continues in this field, there has been and continues to be a great deal of apathy in the UK construction industry regarding the wider consideration of whole-life costs.

It has been estimated that the relationship between capital costs, running costs and business costs in owning a typical office block over a thirty-year period is:

- Construction (capital) cost 1
- Maintenance and operating costs 5
- Business operating costs 200

Although the Royal Academy has recently expressed some doubt as to the accuracy of the ratio, suggesting that in reality it could be closer to 1:2:200, the fact remains that whole-life costs are still a considerable issue in the cost of built asset ownership.

The business operating costs in the above equation (200) include the salaries paid to staff and so on. Clearly in the long term, this aspect is worthy of close attention by design teams and cost advisers. One of the reasons for this lack of interest, particularly in the private sector and the developers' market, is that during the 1980s financial institutions became less enamoured with property as an investment and turned their attention to the stock market. This move led to the emergence of the developer/trader, often an individual, rather than a company, who proposed debt-financed rather than investment-financed development schemes. Whereas previously, development schemes had usually been pre-let and the investor may even have been the end-user, the developer traders had as many projects in the pipeline as they could obtain finance for. The result was an almost complete disregard for whole-life costs as pressure was put on the designers to pare down capital costs at the expense of ultimate performance, since building performance is poorly reflected in rents and value. Fortunately, these sorts of deals have all but disappeared, with a return to the practice of pre-letting and a very different attitude to whole-life costs. If a developer trader was developing a building to sell on they would have little concern for the running costs and so on. However, in order to pre-let a building, tenants must be certain, particularly if they are entering a lease with a full repair and maintenance provision, that there are no 'black holes', in the form of large repair bills, waiting to devour large sums of money at the end of the lease. In the current market therefore sustainability is as important to the developer as the owner/occupier. A building will have a better chance of attracting better quality tenants, throughout its life, if it has been designed using performance requirements across all asset levels, from facility (building), through system (heating and cooling), to component (air-handling unit), and even subcomponent (fans or pumps).

In and around major cities today, it is clear that buildings which attracted good tenants and high rents in the 1980s and early 1990s now tend only to attract secondary or tertiary covenants, in multiple occupancies, leading to lower rents and valuations. This is an example of how long-term funders are seeing their twenty-five to thirty-year investments substantially underperforming in mid-life, thus driving the need for better whole-life-procured buildings.

Whole-life cost procurement includes the consideration of the following factors:

- *Initial* or procurement costs, including design, construction or installation, purchase or leasing, fees and charges.
- *Future* cost of operation, maintenance and repairs, including: management costs such as cleaning and energy costs.
- *Future* replacement costs, including loss of revenue due to non-availability.
- *Future* alteration and adaptation costs including loss of revenue to non-availability.
- *Future* demolition/recycling costs.

Whole-life appraisers may include whatever they deem to be appropriate – provided they observe consistency in any cross-comparisons. The timing of the future costs associated with various alternatives must be decided and then, using a number of techniques described below, assess their impact. Classically, whole-life cost procurement is used to determine whether the choice of, say, a component with a higher initial cost than other like for like alternatives is justified by being offset by reduction of the future costs as listed above. This situation may occur in new-build or refurbishment projects. In addition, whole-life cost procurement may be used to analyse whether in the case of an existing building a proposed change is cost effective when compared with the 'do-nothing' alternative.

There are three principal methods of evaluating whole-life costs:

- Simple aggregation
- Annual equivalent
- Net present value.

Simple aggregation

The basis of whole-life costs is that components or forms of construction that have high initial costs will, over the expected lifespan, prove to be cheaper and hence better value than cheaper alternatives. This method of appraisal involves adding together, without discounting, the initial capital costs, operational and maintenance costs. This approach has a place in the marketing brochure and it helps to illustrate the importance of considering all the costs associated with a particular element but has little value in cost forecasting. A similarly simplistic approach is to evaluate a component on the time required to pay back the investment in a better quality product. For example, where a number of energy-saving devices are available for lift installations, a choice is made on the basis of which, over the life cycle of the lift (say, five or ten years), will pay back the investment the most quickly. This latter approach does have some merit, particularly in situations where the life cycle of the component is relatively short, and the advances in technology and hence the introduction of a new and more efficient product is likely.

Net present value approach

The technique of discounting allows the current prices of materials to be adjusted to take account of the value of money of the life cycle of the product. Discounting is required to adjust the value of costs, or indeed benefits which occur in different time periods so that they may be assessed at a single point in time. This technique is widely used in the public and private sectors as well as sectors other than construction. The choice of the discount rate is critical and can be problematic, as it can alter the outcome of calculation substantially. However, when faced with this problem, the two golden rules that apply are that the public sector follows the recommendations of the Green Book or Appraisal and Evaluation in Central Government, currently recommending a rate of 3.5 per cent. In the private sector the rule is to select a rate that reflects the real return currently being achieved on investments. To help in understanding the discount rate, it may be considered almost as the rate of return required by the investor which includes costs, risks and lost opportunities.

The mathematical expression used to calculate discounted present values is set out below:

$$\text{Present value (PV)} = \frac{1}{(1 + i)^n}$$

where:

(i) = rate of interest expected or discount rate
(n) = the number of years.

This present value multiplier/factor is used to evaluate the present value of sums such as replacement costs that are anticipated or planned at say ten- or fifteen-year intervals.

For example, consider the value of a payment of £150 that is promised to be made in five years' time. Assuming a discount rate of 3.5 per cent, £150 in five years' time would have a current worth or value of £126.30:

$$\frac{1}{(1 + i)^n}$$

$$£150 \times \frac{1}{(1.035)^n} = £150 \times 0.8420 = £126.30.$$

In other words, if £126.30 were to be invested today at 3.5 per cent this sum would be worth £150 in five years' time, ignoring the effects of taxation.

Calculating the present value of the differences between streams of costs and benefits provides the net present value (NPV) of an option and this is used as the basis of comparison as follows.

Annual equivalent approach

This approach is closely aligned to the theory of opportunity costs, i.e. the amount of interest lost by choosing option A or B, as opposed to investing the sum at a given rate percentage, is used as a basis for comparison between alternatives. This approach may also include the provision of a sinking fund in the calculation in order that the costs of replacement are also taken into account.

In using the annual equivalent approach the following equation applies: present value of £1 per annum (sometimes referred to by actuaries as the annuity that £1 will purchase.) This multiplier/factor is used to evaluate the present value of sums, such as running and maintenance costs that are paid on a regular annual basis.

$$\text{Present value of £1 per annum} = \frac{(1 + i)^n - 1}{i\,(1 + i)^n}$$

where (i) = rate of interest expected or discount rate
(n) = the number of years.

Previously calculated figures for both multipliers are readily available for use from publications such as Parry's *Valuation Tables*.

Sinking funds should also be considered, namely a fund created for the future cost of dilapidations and renewals. Given that systems are going to wear out and/or need partial replacement it is thought to be prudent to 'save for a rainy day' by investing capital in a sinking fund to meet the cost of repairs and so on. The sinking fund allowance therefore becomes a further cost to be taken into account during the evaluation process. Whether this approach is adopted will depend on a number of features including corporate policy, interest rates and so on.

Whole-life costing is not an exact science, since, in addition to the difficulties inherent in future cost planning, there are larger issues at stake. It is not just a case of asking 'how much will this building cost me for the next fifty years'; rather it is more difficult to know whether a particular building will be required in fifty years' time at all – especially as the current business horizon for many organisations is much closer to three years. In addition, whole-life costing requires a different way of thinking about cash, assets and cash flow. The traditional capital cost focus has to be altered, and costs thought of in terms of capital and revenue costs coming from the same 'pot'. Many organisations are simply not geared up for this adjustment. The common misconception that a whole-life costed project will always be a project with higher capital costs does not assist this state of affairs. As building services carry a high proportion of the capital cost of most construction projects, this is of particular importance. Just as capital and revenue costs are intrinsically linked, so are all the variables in the financial assessment process. Concentrate on one to the detriment of the others and you are likely to fail.

Perhaps, the most crucial reason is the difficulty in obtaining the appropriate level of information and data.

The lack of available data makes the calculations unreliable. Clift and Bourke (1999) found that despite substantial amounts of research into the development of database structures to take account of performance and WLC there remains a significant absence of standardisation across the construction industry in terms of scope and data available. Ashworth also points out that the forecasting of building life expectancies is a fundamental prerequisite for whole-life cost calculations, an operation that is fraught with problems. While to some extent building life relies on the lives of the individual building components, this may be less critical than at first imagined, since the major structural elements, such as the substructure and the frame, usually have a life far beyond those of the replaceable elements. Clients and users will have theoretical norms of total lifespans but these have often proved to be widely inaccurate in the past. The increased complexity of construction means that it is far more difficult to predict the whole-life cost of built assets. Moreover, if the malfunction of components results in decreased yield or underperformance of the building then this is of concern to the end-user/owner. There is no comprehensive risk analysis of building components available for practitioners, only a wide range of predictions of estimated lifespans and notes on preventive maintenance. This is too simplistic; there is a need for costs to be tied to risk including the consequences of component failure. After all the performance of a material or component can be affected by such diverse factors as:

- Quality of initial workmanship when installed on site and subsequent maintenance.
- Maintenance regime/wear and tear. Buildings that are allowed to fall into disrepair prior to any routine maintenance being carried out will have a different life cycle profile to buildings that are regularly maintained from the outset.
- Intelligence of the design and the suitability of the material/component for its usage. There is no guarantee that the selection of so-called high-quality materials will result in low life cycle costs.

Other commonly voiced criticisms of whole-life cost are:

- Expenditure on running costs is 100 per cent allowable revenue expense against liability for tax and as such is very valuable. There is also a lack of taxation incentive in the form of tax breaks and so on for owners to install energy-efficient systems.

  In the short term, and taking into account the effects of discounting the impact upon future expenditure is much less significant in the development appraisal.
- Another difficulty is the need to be able to forecast, a long way ahead in time, many factors such as life cycles, future operating and maintenance costs, and discount and inflation rates. WLC, by definition, deals with the future, and the future is unknown. Increasingly obsolescence is being taken into account during procurement, a factor that it is impossible to control since it is influenced by such things as fashion, technological advances and innovation. An increasing challenge is to procure built assets with the flexibility to cope with changes. Thus, the treatment of uncertainty in information and data is crucial, as uncertainty is endemic to WLC (Flanagan *et al.*, 1989; Bull, 1993). Another major difficulty is that the WLC technique is expensive in terms of the time required. This difficulty becomes even clearer when it is required to undertake a WLC exercise within an integrated real-time environment at the design stage of projects.

In addition to the above, changes in the nature of the development of other factors have emerged to convince the industry that whole-life costs are important.

Whole-life cost procurement – critical success factors:

- Effective risk assessment – what if this alternative form of construction is used?
- Timing – begin to assess WLC as early as possible in the procurement process.
- Disposal strategy – is the asset to be owner occupied, sold or let?

- Opportunity cost – downtime.
- Maintenance strategy/frequency – does one exist?
- Suitability – matching a client's corporate or individual strategy to procurement.

Sources of cost data for whole-life cost calculations

Kelly and Hunter (2005) and Flanagan and Jewel (2005) cite the basic data sources as:

- data from specialist manufacturers, suppliers and contractors;
- predictive calculations from model building;
- historic data.

Flanagan and Jewel highlight the danger associated with data used for life cycle costing, stating that:

- Data are often missing.
- Data can often be inaccurate.
- People often believe they have more data than actually exists.
- It can be difficult to download data for subsequent analyses and for data sharing by a third party.
- There will be huge variation in the data, sometimes for the same item.
- Data are often not up to date.
- Data input is unreliable: the input should be undertaken by those with a vested interest in getting it right.

Both Kelly and Hunter and Flanagan and Jewel quote the UK Office of Government Commerce (2003) which states that it is important to focus on future trends rather than compare costs of the past. Where historic data are available they may provide misleading information, such as past mistakes in the industry and focusing on lowest price. Historic data are best used for budget estimates at whole building or elemental levels. At the point of option appraisal of systems and components it is always preferable to estimate the cost from first principles and only to use historical cost information as a check.

Whole Life Cost Forum

In 2001 the Whole Life Cost Forum was launched as a source of reference and historical whole-life cost data and may be accessed at http//:www.wlcf.org.uk.

Building Maintenance Information (BMI)

Building Maintenance Information (BMI) has recorded the cost of occupying buildings in the UK for over thirty years, and has collected data on the occupancy and maintenance costs of buildings from subscribers and other sources. This service has been relaunched as BCIS Building Running Costs Online and is a web-based service for professionals involved in facilities management, maintenance and refurbishment. A central database is organised in an elemental format allowing comparative analyses to be undertaken, rebased for time and location based upon indices updated monthly. The service also keeps life expectancy of building components data. BCIS Running Costs Online has a life cycle costing module that combines the information from the BCIS annual reviews of maintenance and occupancy costs with the data from the biannual occupancy cost plans, allowing users to compare the running costs of different building types. The output is a spend profile over a period of up to sixty years showing the estimated expenditure for each year of the selected period.

There are a number of definitions for whole-life costing, but one currently adopted is: 'the systematic consideration of all relevant costs and revenues associated with the acquisition and ownership of an asset' (Figure 3.5).

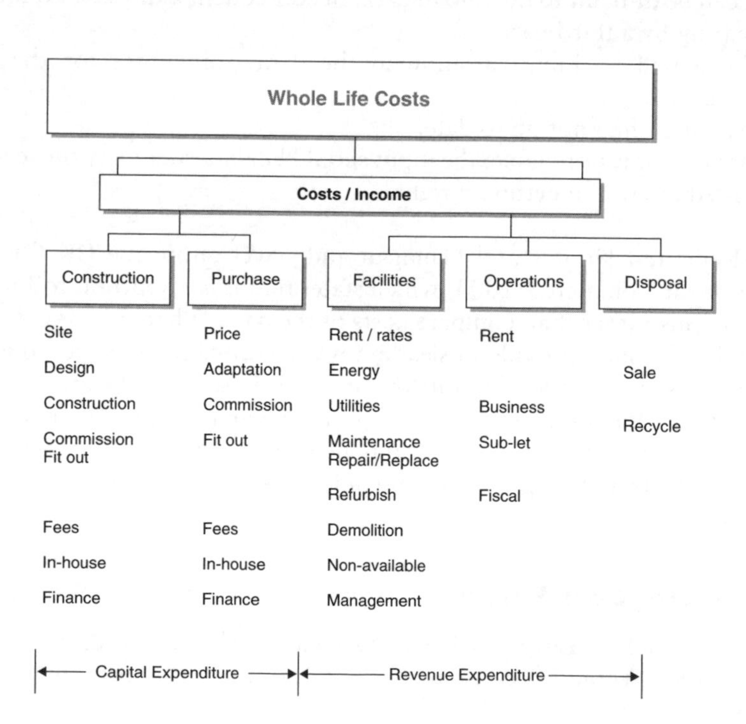

Figure 3.5 Whole-life costs

Although whole-life costing can be carried out at any stage of the project and not just during the procurement process, the potential of its greatest effectiveness is during procurement because:

- almost all options are open to consideration at this time;
- the ability to influence cost decreases continually as the project progresses, from 100 per cent at project sanction to 20 per cent or less by the time construction starts;
- the decision to own a building normally commits the user to most of the total cost of ownership and consequently there is a very slim chance to change the total cost of ownership once the building is delivered.

Typically, about 75–95 per cent of the cost of running, maintaining and repairing a building is determined during the procurement stage.

There now follows a simple example, based on the selection of material types, illustrating the net present value and the annual equivalent approaches to whole-life cost procurement (Table 3.5).

This problem is a classic one. Which material, with widely varying initial and maintenance costs, will deliver the best value for money over the life cycle of the building? In this example, and supposing a discount rate of 6 per cent, it is assumed that the materials are to be considered for installation in a PFI project, with an anticipated life of twenty-five years.

Table 3.6 indicates a whole-life cost calculation for material A presented in two ways: as a net present value and as an annual equivalent cost. The calculation is repeated for each material or component under consideration and then a comparison made.

A replacement expenditure profile, excluding cyclical maintenance and energy over a range of elements over a thirty-five-year contract period, is shown in Table 3.6.

Clearly, the choice of the correct type of material or component would appear to be of critical importance to a client, as future replacement and maintenance costs will have to be met out of future income. However, in reality theory and practice are often very different. For example, for many

Table 3.5 Material costs

Material	*Initial cost*	*Installation cost*	*Maintenance cost per day*	*Other maintenance costs*	*Life expectancy*
A	£275	£150	£3	£100 every 3 years for preservative treatment	12 Years
B	£340	£150	£3	None	15 Years

Table 3.6 Results for Material A

Year	*Present value of £1 per annum (PV of £1 pa)*	*Present value (PV £1)*	*Initial cost £*	*Other costs £*	*Annual cost £ £3 × 365*	*Total Discounted Costs NPV of replacement + other + annual costs + initial costs £*	*Total NPV £*	*AEC £*
1	0.943	0.943	490.00		1095.00	1523.02	1523.02	1614.40
2	1.834	0.890			1095.00	974.55	2497.57	1362.25
3	2.673	0.840			1095.00	919.38	3416.95	1278.31
4	3.465	0.792			1095.00	897.34	4284.29	1236.41
5	4.212	0.747			1095.00	818.25	5102.54	1211.32
6	4.917	0.705			1095.00	771.93	5874.47	1194.65
7	5.582	0.665			1095.00	728.24	6602.71	1182.76
8	6.210	0.627			1095.00	687.02	7289.72	1173.91
9	6.802	0.592			1095.00	648.13	7937.85	1167.04
10	7.360	0.558			1095.00	611.44	8549.30	1161.58
11	7.887	0.527			1095.00	576.83	9126.13	1157.13
12	8.384	0.497			1095.00	544.18	9670.31	1153.47
13	8.853	0.469			1095.00	513.38	10183.69	1150.35
14	9.295	0.442			1095.00	484.32	10668.00	1147.72
15	9.712	0.417		490.00	1095.00	661.37	11329.38	1166.50
16	10.106	0.394			1095.00	431.04	11760.42	1163.72
17	10.477	0.371			1095.00	406.64	12167.06	1161.28
18	10.828	0.350			1095.00	383.63	12550.69	1159.14
19	11.158	0.331			1095.00	361.91	12912.60	1157.24
20	11.470	0.312			1095.00	341.43	13254.02	1155.55
21	11.764	0.294			1095.00	322.10	13576.12	1154.03
22	12.042	0.278			1095.00	303.87	13879.99	1152.67
23	12.303	0.262			1095.00	286.67	14166.66	1151.45
24	12.550	0.247			1095.00	270.44	14237.10	1150.33
25	12.783	0.233			1095.00	255.13	14692.24	1149.36

Notes
AEC = Annual equivalent cost.
Other cost = replacement costs every 15 years.

Table 3.7 A replacement expenditure profile over thirty-five years

Element	*Replacement expenditure (%)*
Windows/doors	22.95
Kitchens	15.79
Heating	11.82
Structural	10.63
Roofs	8.72
Bathrooms	7.79
Wiring	6.50
External areas	3.87
Internal decorations	2.48
Communal decorations	1.69
Over cladding	1.61
Rainwater goods	1.51
External walls	0.99
Off-road parking	0.82
DPC	0.73
Security/CCTV	0.60
Door entry systems	0.51
Fire precautionary works	0.50
Porches/canopies	0.44
Plastering	0.07

Source: Whole life costs forum.

public authorities, finding budgets for construction works is usually more difficult than meeting recurring running and maintenance costs that are usually included in annual budgets as a matter of course.

In addition to the net present value and annual equivalent approaches described previously, Williams identified that simple aggregation could sometimes be used effectively when evaluating whole-life costs.

Costs of alternative (renewable) materials

Renewable construction materials are made from plant-derived substances that can be produced repeatedly. In contrast, most other construction materials are derived from raw materials which we cannot replace: oil, minerals, metals and so on. Renewable construction materials range from very 'natural' unprocessed materials such as straw bales used for walling, to more refined products such as floor coverings manufactured using renewable polymers. Renewable materials are beginning to be introduced into mainstream construction (e.g. hemp and lime walling material), and it is claimed may soon become commonplace in UK buildings.

Renewable construction materials require less embodied energy to manufacture, process and transport them to their point of use. Embodied energy includes energy used in obtaining the raw materials (e.g. sand and gravel or plant materials), energy used in the processing of those materials (e.g. grinding, blending, firing), and energy used in the manufacture of finished products. Embodied energy may be regarded as an energy 'debt' incurred by materials; therefore, using materials with low embodied energy reduces the carbon footprint of a building and reduces its impact upon the environment.

The use of renewable construction materials is not a requirement in building regulations and is not required for ratings to be awarded in the Code for Sustainable Homes. This may be partially due to the fact that renewable construction materials are still in development in the UK and are not yet widely available for most applications. For these reasons information on costs associated with renewable materials is not widely available.

Renewable energy schemes

Renewable energy is defined as energy flows which are replenished at the same rate as they are used. Renewable energy may be direct; for example, solar water heating, or indirect; for example, biomass, wind and hydro. Renewable energy schemes are not without their critics, the most common criticism being the sheer scale of, say, wind farm schemes that will be required to replace conventional power generation. Another point of contention is that with the extra cost of installation (for example, the cost of a large woodchip-burning biomass boiler) the pay-back period can be considerable, perhaps as much as 100 years-plus; in addition, the extra space required for fuel storage could be considerable. For example, the fuel demand of a 20,000kWh boiler over a heating season equates to 2m^3 of kerosene as against 25m^3 of woodchips. In addition, there is the necessity to source fuel from within a twenty-five-mile radius of the boiler in order to maintain a satisfactory carbon footprint. Against this the running cost of woodchip is substantially below the cost of conventional fuels, perhaps half in terms of cost per kWh. At the time of writing the cost information biomass installation available to the quantity surveyor is limited.

Perhaps the most widely used applications of renewable energy sources are solar and photovoltaic panels. Used most often for domestic scale schemes the main limiting factor is the area of panels required to produce energy. For example, in the case of solar panels 1m^2 of panel is required to deliver 45 litres of hot water at a cost of approximately £750 to £800 per m^2 to install. The payback period for solar panels and photovoltaic panels is again considerable and could conceivably be longer than the design life of the project. Planning consideration may also be an issue.

For many people the most obvious signs of renewable energy schemes are wind turbines and wind farms. The costs of installation are considerable, but these are capable of being off-set if the electricity produced is suitable for connection to the national grid; for example, £40,000 for an entry level turbine capable of producing 15kWh. Other considerations are tax allowances and of course the suitability of the site for locating a wind turbine.

Ska ratings

Formally launched on 6 November 2009, the Ska rating is an environmental assessment method designed to rate and compare the environmental performance of fit-out projects, initially for office buildings in the UK. Ska ratings are similar to BREEAM credits but focus solely on assessing a fit-out. In the case of a BREEAM Offices fit-out assessment, the tool filters out the land use and ecology credits as well as some of the credits relating to the build construction, therefore tailoring it for a fit-out assessment. However, some of the BREEAM Offices fit-out assessment credits do relate to the building which can be outside the control of the project (e.g. whether the building has a pulsed water meter is assessed in BREEAM but the Ska rating only considers a water meter if the provision or modification of water services is within the scope of the fit-out). In addition, the BREEAM fit-out assessment has credits relating to the proximity of the building to public transport nodes (Ska rating does not assess this).

The Ska assessment process is broken down into three stages:

- *Design/planning*. At this stage we identify the measures and issues in scope. Once the measures in scope are identified the client has the opportunity to prioritise which measures they want to achieve and to make a decision against design, cost, programme and benefit, and add them into the scope of works for the project. This will also set the environmental performance standards for how the project is delivered, in terms of waste and energy in use and so on. Then if the specification demonstrates that these measures are likely to be achieved they will be reflected in an indicative rating.
- *Delivery/construction*. This involves the gathering of evidence from O&M manuals and other sources to prove that what has been specified has actually been delivered and that the performance and waste benchmarks have been achieved.
- *Post-occupancy assessment*. Finally there is the option to review how well a fit-out has performed in use against its original brief from a year after completion.

The RICS is currently operating an accreditation scheme for Ska assessment. The RICS charges £50 for each certification an assessor carries out. The assessor could typically charge £2,000 to £3,000 per certification, depending on the nature and complexity of the project.

Recycled contents schemes

Recycled content is defined in ISO 14021:

> Recycled content is the proportion, by mass, of recycled material in a product or packaging. Only pre-consumer and post-consumer materials shall be considered as recycled content, consistent with the following usage of the terms:
>
> - Pre-consumer material: material diverted from the waste stream during a manufacturing process. Excluded is reutilization of materials such as rework, regrind or scrap generated in a process and capable of being reclaimed within the same process that generated it.
> - Post-consumer material: material generated by households or by commercial, industrial and institutional facilities in their role as end-users of the product, which can no longer be used for its intended purpose. This includes returns of material from the distribution chain.

The use of recycled materials is already a requirement for a number of construction clients as follows:

- The Scottish government has asked all public bodies in Scotland to set 10 per cent recycled content as a minimum standard in major public sector projects in Scotland. Councils including Aberdeen, Glasgow, Midlothian, South Ayrshire and the Shetland Islands have already taken action, as has Scottish Water.
- The Central Procurement Directorate in Northern Ireland issued recycled content guidance in February 2006.
- The Welsh Assembly government has set a 10 per cent recycled content target in major regeneration projects and Welsh Health Estates applies a KPI and target in health sector procurement.
- The Olympic Delivery Authority has adopted minimum standards of at least 20 per cent (by value) of materials used in the permanent venues, to be from recycled content for London 2012.
- In England, the late Building Schools for the Future programme, Defence Estates, the National Offender Management Service and hospital PFI projects such as Bristol South Mead and Hillingdon have all adopted KPIs and benchmarks for recycled content.
- Property developers and retailers including British Land, Hammerson, John Lewis Partnership, Marks and Spencer and Stanhope have also set recycled content targets.
- Councils including Bristol, Greenwich, Islington, Lancashire, Leeds, Newcastle, Nottingham, Sandwell and Sheffield have set recycled

content tender requirements in schools PFI, as have Leeds Metropolitan and Worcester Universities.
- Minimum recycled content standards have been adopted for regeneration by South West England and Yorkshire Forward Regional Development Agencies, Leeds Holbeck and Raploch Urban Regeneration Company.

According to the Waste and Resources Action Programme (WRAP) Table 3.8 gives examples of higher recycled content products available at no extra cost.

Material types which are believed to offer higher levels of recycled content are given in Table 3.9.

The ethical sourcing of materials and labour

According to ICLEI key elements of ethical sourcing include:

- Equal partnership and respect between producers and consumers
- A fair price for socially just and environmentally sound work
- Healthy working conditions
- Fair market access for poverty alleviation and sustainable development
- Stable, transparent and long-term partnership
- Guaranteed minimum wages and prompt payment
- Financial assistance, when needed (pre-production financing)

Table 3.8 Examples of higher recycled content products

Component	*Typical product*	*Products with higher recycled content*
Dense blocks	Brand A – 0% recycled content (£5.50/m²)	Brand B – 50–80% recycled content (£5.50/m²)
Concrete roof tiles	Brand C – 0% recycled content (£550/1000)	Brand D – 25% recycled content (£550/1000)
Glass/mineral wool insulation	Brand E – 10% recycled content (£3.50/m²)	Brand F – 80% recycled content (£3.00m²)

Source: Wrap – Material change for a better environment.

Table 3.9 Material types offering higher levels of recycled content

Bulk aggregates	Ready-mix concrete
Asphalt	Drainage products
Pre-cast concrete products	Concrete tiles
Clay facing bricks	Lightweight blocks
Dense blocks	Plasterboard
Ceiling tiles	Chipboard
Insulation	Floor coverings

- Premiums on products used to develop community projects
- Encouraging better environmental practices

Advice is needed by all types of owners, occupiers, lenders, investors and public and private bodies as to:

- their environmental duties and liabilities;
- how to determine and quantify liability;
- the implications for asset management arising from any actual or potential liabilities;
- who to look to for advice and how advisers should be appointed;
- the steps to take to minimise or eliminate liability;
- the likelihood of ongoing, new or potential liability.

The business case for green development

There now seems to be a greater awareness that energy-efficient offices, for example, command higher rental prices, have lower vacancy rates and higher market values, relative to otherwise comparable conventional office buildings. This, along with revenue savings of reduced energy and water as well as expert opinion that suggests higher levels of productivity can be achieved, begins to create a compelling argument for building and refurbishing to higher green standards. However, despite this, energy-efficient buildings are still in the minority of total building stock.

Comparatively few green buildings have been completed, and of those a high percentage has been in the public sector. However, proponents of green development maintain that sustainable buildings can:

- Significantly reduce whole-life costs and ensure more rapid pay-back compared to conventional buildings from lower operation and maintenance costs, thereby generating a higher return on investment.
- Secure tenants more quickly.
- Command higher rents or prices.
- Enjoy lower tenant turnover.
- Attract grants, subsidies, tax breaks and other inducements.
- Improve business productivity for occupants.
- Eighty-five per cent of a building's real costs are related to staff/productivity costs – user satisfaction is therefore key.
- Initial construction costs <10 per cent of a building's lifetime costs.
- Energy is typically 30 per cent of a building's operating costs.
- Increase productivity especially through day lighting.
- Promote enhanced health and well-being.
- Produce higher academic achievement.
- Induce higher morale in staff.

- Promote reduced absenteeism.
- Enhance image – branding and symbolising values.

A raft of new legislation is now ensuring that the consideration of sustainability for new projects is not merely an option. In January 2006 the European Commission's Energy Performance Directive came into effect, followed shortly in April 2006 by the Code for Sustainable Buildings. If this were not enough the long-awaited revision to Part L of the Building Regulations is also now in force. The Building Regulations are considered to be one of the most far-reaching pieces of legislation ever to hit the construction industry and will force cuts in carbon emissions from buildings by one million tonnes a year. Design teams will have to obtain energy ratings before and after construction and assessment will be based on the government's Standard Assessment Procedure for Energy Rating 2005 (SAP, 2005).

Other considerations

Off-site construction

The benefits of off-site factory production in the construction industry are well documented and include the potential to considerably reduce waste, especially when factory-manufactured elements and components are used extensively. Its application also has the potential to significantly change operations on site, reducing the amount of trade and site activities and changing the construction process into one of a rapid assembly of parts that can provide many environmental, commercial and social benefits, including:

- reduced construction-related transport movements;
- improved health and safety on site through avoidance of accidents;
- improved workmanship quality and reducing on-site errors and rework, which themselves cause considerable on-site waste, delay and disruption;
- reduced construction timescales and improved programmes.

Off-site construction is one of a group of approaches to more efficient construction, sometimes called modern methods of construction, that also include prefabrication, improved supply chain management and other approaches. Technologies used for off-site manufacture and prefabrication range from modern timber and light gauge steel framing systems, tunnel form concrete casting through to modular and volumetric forms of construction, and offer great potential for improvements to the efficiency and effectiveness of UK construction.

Finally, key questions to consider during the procurement stages are:

- Has research been carried out by the design team and/or use of the WRAP Net Waste Tool to identify where on-site waste arises?
- Can construction methods that reduce waste be devised through liaison with the contractor and specialist subcontractors?
- Have specialist contractors been consulted on how to reduce waste in the supply chain?
- Have the project specifications been reviewed to select elements/components/materials and construction processes that reduce waste?
- Is the design adaptable for a variety of purposes during its lifespan?
- Can building elements and components be maintained, upgraded or replaced without creating waste?
- Does the design incorporate reusable/recyclable components and materials?
- Are the building elements/components/materials easily disassembled?
- Can a Building Information Modelling (BIM) system or building handbook be used to record which and how elements/components/materials have been designed for disassembly?

Bibliography

Dixon, T. *et al.* (2007). *Green Profession?: An Audit of Sustainability Tools, Techniques and Information for RICS Members*. RICS, London.

Flanagan, R. and Jewel, C. (2005). *Whole Life Appraisal for Construction*. Blackwell, Oxford.

HM government (2008). *Strategy for Sustainable Construction*. Department for Business Enterprise and Regulatory Reform.

IPF Research (2009). *Costing Energy Efficiency Improvements in Existing Commercial Buildings*. Investment Property Forum.

Kelly, J. and Hunter, K. (2005). *A Framework for Whole Life Costing*. SCQS.

Kelly, J. and Hunter, K. (2009). *Life Cycle Costing of Sustainable Design*. RICS Research Report.

RICS (2007). *Surveying Sustainability; A Short Guide for the Property Profession*. RICS, London

RICS (2009a). *Sustainability and the RICS Property Life Cycle*. RICS, London.

RICS (2009b). *Renewable Energy*. RICS, London.

Secretary of State for Energy and Climate Change (2009a). *The UK Renewable Energy Strategy*. HMSO.

Secretary of State for Energy and Climate Change (2009b). *The Low Carbon Transition Plan*. HMSO.

Williams, B. (2006). *Benchmarking of Construction Costs in the Member States*. BWA.

Wrap, Cyril Sweett (2009). *Delivering Higher Recycled Content in Construction Projects*. Waste Resources Action Programme.

Chapter 4

IT update

The first edition of *New Aspects of Quantity Surveying Practice* was drafted in what was for the IT and e-commerce sectors the heady days of the late 1990s/early 2000. There were exaggerated predictions from practically every quarter, including government departments, on the ways by which e-commerce and e-construction were going to become all-pervading and revolutionise everything from tendering to project management. To paraphrase Harold Wilson's famous 1960s speech where he referred to the 'white heat of technology', the late 1990s and the early part of this new decade was a period when the 'white heat of information technology' resulted in a massive expansion of e-commerce portals and its associated systems.

The reality, as described in thc following paragraphs, has been somewhat different! Nevertheless, the perception of e-commerce by construction-related organisations and professions is mostly positive, although to date objective measures have been missing to accurately benchmark the spread of e-construction. There is now a universal consensus that reliable e-commerce metrics are needed to track developments in this area and to understand its impact upon our economies and societies. The OECD identifies the research and measurement priorities as follows:

Readiness:	Potential usage and access Technology infrastructure/socio-economic infrastructure
Intensity:	Transaction/business size Nature of transaction/business
Impact:	Efficiency gains Employment/skill composition; work organisation New products/services New business models Contribution to wealth creation Changes in product/value chains

E-commerce defined

E-commerce is a major business innovation which tends to be successful when led by commercial rather than technological considerations. E-commerce exploits information and communication technologies (ICTs) to re-engineer processes along an organisation's value chain in order to lower costs, improve efficiency and productivity, shorten lead-in times and provide better customer service. Electronic commerce or e-commerce therefore consists of the buying, selling, marketing and servicing of products or services over computer networks. Strictly speaking, according to the OECD, e-commerce may be defined only as the method by which an order is placed or received and does not extend to the method of payment or channel of delivery; therefore for the purist, the simple process of gathering information online or sending an e-mail does not constitute e-commerce. Whatever the arguments over the definition, for the quantity surveyor, e-commerce allows for instant communication through the supply chain, giving the partners a clear real-time picture of supply and demand.

One of the first organisations to use the term e-commerce was IBM, in October 1997 when it launched a thematic campaign built around the term. Over the past few years major corporations have re-engineered their business in terms of the internet and its new culture and capabilities, and construction and surveying have started to follow this trend. As with any new innovation there are forces that act to drive forward the new ideas as well as those forces that act as inhibitors to progress. These are illustrated in Figure 4.1 and discussed later in the chapter. E-commerce sprang to public attention in 1997, after the meteoric rise in the value of the so-called dot.com companies. Many within the construction industry were sceptical about the application of e-commerce to construction and in 2000/2001 when the value of shares on the UK stock market collapsed and many virtual companies evaporated overnight, leaving massive debts and red faces, there was a collective 'I told you so!'

The general consensus is that the failure of many dot.com companies occurred owing to a combination of some or all of the following factors:

- Poor business models;
- Poor management;
- Aggressive spending;
  Forgetting the customer;
- The changing attitudes of venture capitalists.

In 1999/2000 alone it was estimated that the total amount of venture capital investment in B2B (see later definition) exchanges was $25 billion worldwide, but this decreased to a trickle by 2001.

In common with other market sectors, e-construction has had a number of false dawns; for example, at the height of dot.com mania five of the largest contractors in the UK announced the creation of the first industry-wide electronic marketplace offering the purchase of building materials online as well as a project collaboration package. Just over a year later it was announced that the planned internet portal was to be shelved, due to lack of interest. Despite this lack of enthusiasm from within the industry and the professions, due in large part to the continuing client lead drive for efficiency and added value, as well as examples from other sectors, the move to become an e-enabled industry continues. Information technology is at the heart of the developing tools and technologies that pool information into databases. An equally important aspect is looking at the attitudes of the people who need to feed information into and to use the system. The internet in particular provides a platform for changing relationships among clients, surveyors, contractors and suppliers; open exchange of information is critical in order to harness the best from this virtual marketplace and one of the biggest challenges is creating a culture that encourages and rewards the sharing of information. Too many people are still starting from the viewpoint that knowledge is power and commercial advantage, and the belief that the more they keep the knowledge to themselves, the more they will be protecting their power and position. Despite the somewhat slower uptake by the construction industry compared with other sectors, Figure 4.1 illustrates the wide range of e-markets available for construction and quantity surveying applications from basic e-shop websites to complex value chain integrators and collaboration platforms. Why then has e-commerce not become the dominant way of conducting business in the construction sector? In a survey of the construction industry commissioned by the Construction Products Association in 2005, 83 per cent of those surveyed cited the culture of the industry as the major constraint on the development of e-commerce in construction. The survey also concluded that at first sight little had changed since 2000 when 86 per cent of those questioned in a similar survey cited the same cause for the slow uptake of e-commerce. Yet, despite this negativity, 80 per cent believed that the construction industry is still committed to embracing new technologies; however, growth at 26 per cent had been much slower than the predicted 50 per cent in 2001.

The key advantages for the adoption of e-technologies were considered, by respondents in the 2005 survey, to be:

- reducing costs;
- faster transactions;
- fewer errors;
- reduced paperwork.

Interestingly, 'access to new markets' which ranked very highly in the early days of e-commerce surveys has now disappeared from the list of perceived advantages.

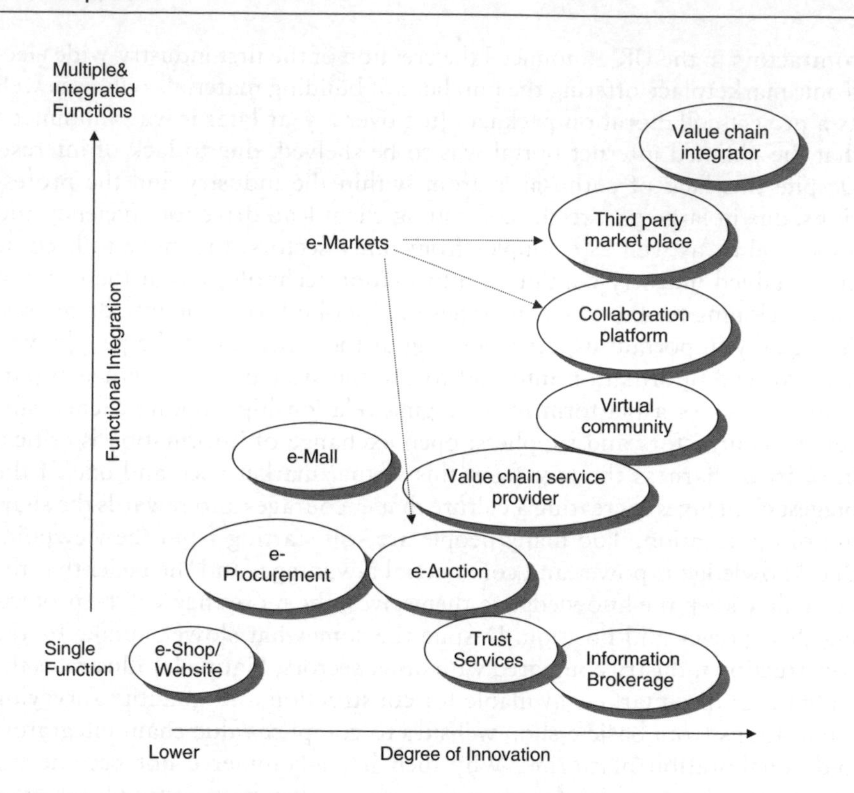

Figure 4.1 Classification of Electronic Commerce Business Models
Source: Paul Timmers.

The main disadvantages of e-technologies were seen to be:

- initial set-up costs;
- loss of personal contact;
- retraining of staff.

One of the more surprising results of the Construction Products Association survey was respondents' views on the effect of e-commerce on construction products. While 80 per cent of the industry as a whole and 69 per cent of manufacturers and distributors expected little or no change in the range of products available, 10 per cent of the industry as a whole and 19 per cent of construction product manufacturers expected a decrease in product awareness. Looking forward to 2012, when asked to predict how the proportion of business done with their suppliers and customers using e-commerce would change, the highest increase was envisaged to be with suppliers, with an overall increase of 48 per cent compared to 30 per cent with customers. The above figures indicate that to date, the majority of organisations would appear to use e-commerce to track the competition and improve communications. In addition, it would appear that many companies, particularly SMEs, are engaging in e-commerce activities as a result of competitive pressure, suggesting a defensive line of action rather than a differentiated one. However, as clients become more e-enabled, quantity surveying practices must follow or be left behind. As in the case of supply chain management techniques, and certain public sector procurement agencies, the pressure will come from the client; even so, few if any quantity surveying practices have made the leap from simple website to transaction platforms.

A high percentage of quantity surveying practices now have their own websites. Software such as Microsoft Expression Web, as well as inexpensive proprietary website templates makes the process of producing a professional-looking website comparatively simple and inexpensive (e.g. www.duncancartlidge.co.uk). This so-called first generation presence is used mainly for marketing and employs the simplest form of business model (illustrated in Figure 4.1). The so-called second and third generation presence, which incorporates transaction applications, has shown a much slower growth rate, particularly in the construction sector.

In a survey of e-commerce adoption carried out by Statistical Indicators Benchmarking the Information Society (SIBIS) across seven EU states it was found that the leading e-commerce all-rounders were in distribution and financial network sectors, whereas offline and basic online organisations were most likely in construction and manufacturing.

As shown in Figure 4.2, internet technologies may be exploited in marketing and sales by introducing web marketing and eventually e-sales; this is referred to as the front office development path of e-commerce, since it involves dealing online with final customers. Integration of closed business networks involving suppliers and distribution networks is defined as the back office development path. The next step is the integration of applications and exploiting processes synergies – the all-round e-commerce model.

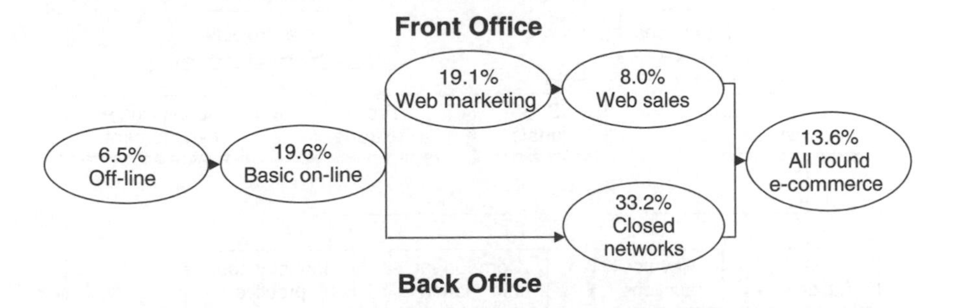

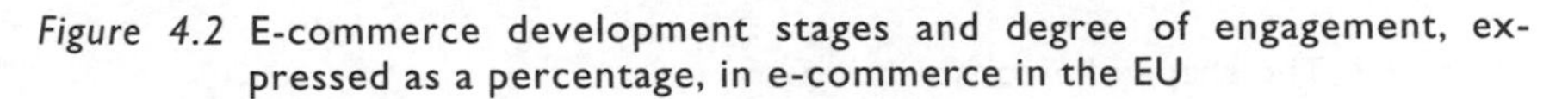

Figure 4.2 E-commerce development stages and degree of engagement, expressed as a percentage, in e-commerce in the EU

Source: empirica SIBIS.

The transparency of the internet should be a driving force for changing business strategies and attitudes and yet it will take a quantum leap in construction business culture to disclose sensitive information to the supply chain. It has been suggested by a leading construction industry dot.com that the European construction industry could save up to 175 billion euros per annum on building costs and reduce completion time by up to 15 per cent through the widespread adoption of e-construction technologies.

When the lean thinking initiative was introduced into construction, it was the car industry that provided the role model. Now that some sections of the construction industry are talking seriously about e-commerce, it can once again look to the motor industry for a lead. In America, the three major domestic motor manufacturing firms have been dealing with suppliers via a single e-commerce site for a number of years – an initiative that has resulted in reported savings of over £600 million a year. Similar initiatives are also to be found in the retail and agriculture sectors, but perhaps in the rush to establish the first truly successful construction-based e-portal the UK players ignored some basic business rules.

Several years after the advent of e-construction the current shape may be considered as follows (and see Figure 4.3).

The shape of e-construction

E-resources; B2B portals

B2B portals combine a number of easily accessible e-markets and include the following resources:

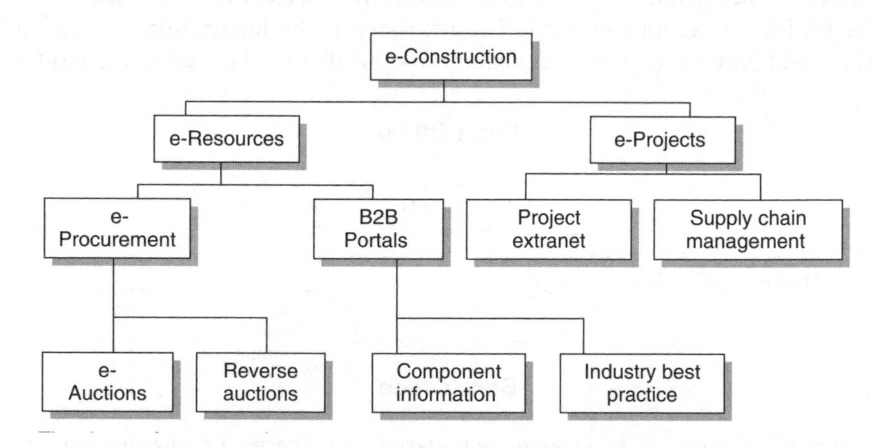

Figure 4.3 The shape of e-construction

Source: Adapted from Winch.

- industry best practice (e.g. www.constructingexcellence.org.uk);
- material and component information.

E-portals can be broken down into categories, based on who is trading with whom. Latterly, sectors such as construction and finance have coined their own terms (e.g. e-construction, e-finance) to stake their own unique claim in the electronic marketplace. Although some sources claim that there can be up to nine classifications of e-commerce, most people agree that there are only three, and of these only two – Business to Business (B2B) and Business to Consumer (B2C) – have seen strong growth, the other sector being Consumer to Consumer (C2C) or Person to Person (P2P).

> *Business-to-Business (B2B) or Business to Administration (B2A); for example, online data exchange.* The extent to which B2B has been adopted by business depends on the sector and the size of the organisation. This category of e-business uses the internet and extranets and it is forecast that spending in B2B is expected to dominate internet growth until 2012. There are basically two different types of B2B companies: horizontal and vertical. As illustrated in Figure 4.5 vertical B2B companies work within an industry and typically make money from advertising on specialised sector-specific sites or from transaction fees from the e-commerce that they may host. Horizontal B2B companies are a completely different breed and operate at different levels across numerous different verticals. Whether it is enabling companies to electronically procure goods, helping to make manufacturing processes run more efficiently, or empowering sales forces with critical information, most horizontal companies make their money by selling software and related services (see e.g. www.tendersontheweb.com).

The development of B2B commerce has been rapid compared with other technological innovations. Figure 4.4 illustrates the development of B2B commerce. Following the establishment of closed EDI networks the second phase of e-commerce saw the emergence of one-to-one selling from websites and a few early adopters began pushing their websites as a primary sales channel (e.g. Cisco and Dell). Most of the initial websites were not able to process orders or to supply order tracking. The current phase of development is represented by the butterfly model and involves the rise of third-party sites that bring together trading partners into a common community. Therefore, in this situation buyers and sellers start to visit the site regularly and move to the collaboration phase, which creates the opportunity to serve a large percentage of those interests. In a few years B2B has developed from a high-cost, rigid system with low transparency to a low-cost, highly flexible and fully transparent system.

The development of e-commerce towards e-business, the more comprehensive definition which is now preferred, reveals the move from the

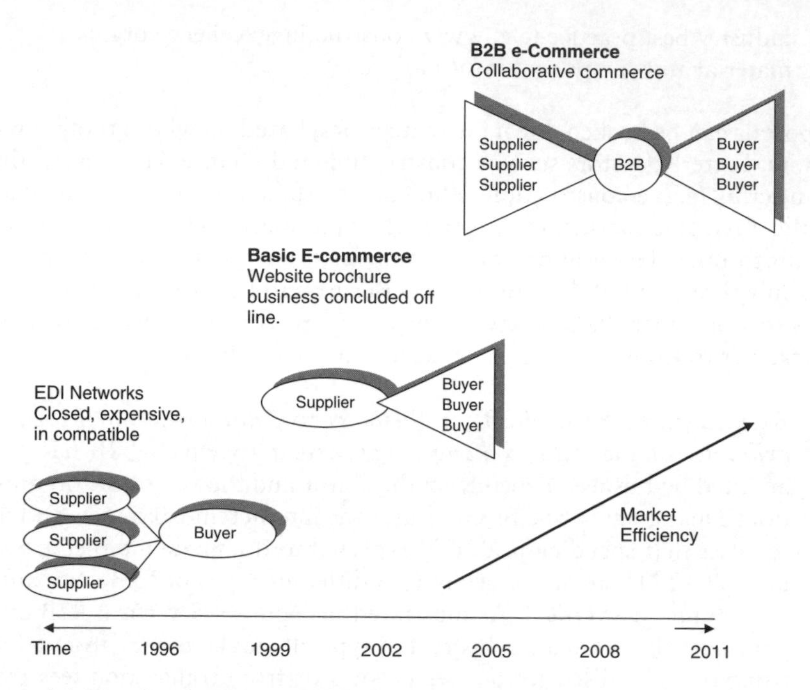

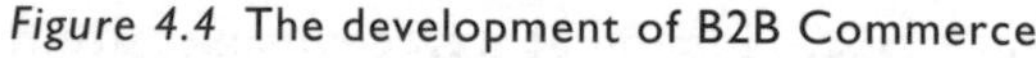

Figure 4.4 The development of B2B Commerce

Source: Morgan Stanley Dean Witter Internet Research.

enterprise-centric vision to the multi-enterprise network or virtual enterprise, and the move towards the exploitation of ICT and the potential of the internet from cost-reduction-oriented e-commerce to the collaborative-commerce vision, a move that requires a major step change in business culture (Figure 4.4).

The challenge for construction and the surveying professions is undoubtedly greater; Dell for example, has the considerable advantage of being already in the internet-related product field, where clients and suppliers are already technologically sophisticated. Nevertheless, real value-added benefits are available to the quantity surveyor. The key point which is now currently acknowledged is that with the use of ICTs and the internet in business, costs can be reduced and value created. Value is created from brokering transactions and matching orders between companies, but also from the provision of additional services, such as professional services for integrating and managing companies including, legal and financial services, logistics, project and contract management as well as background services such as market intelligence. For example, Dell has created virtual integration with both its upstream partners and its downstream clients so that the entire supply chain acts like a single, integrated company. Dell builds computers to order; typically someone who works for a large company like Boeing can

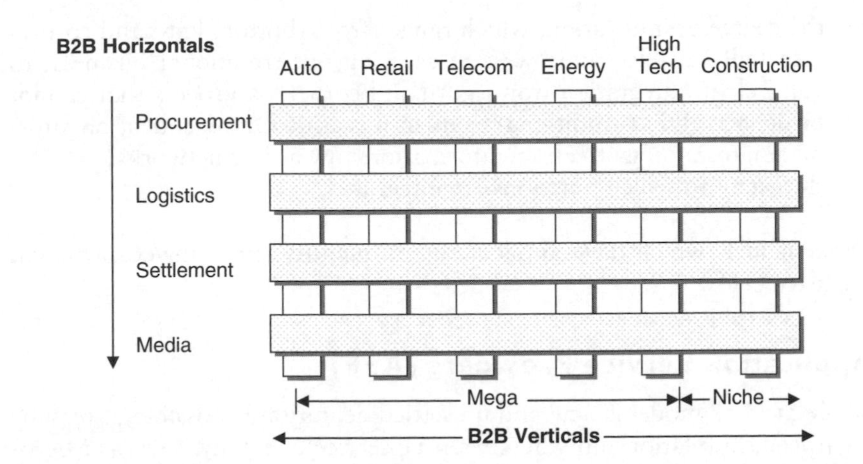

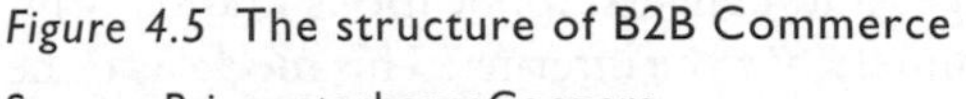
Figure 4.5 The structure of B2B Commerce
Source: PricewaterhouseCoopers.

go to a private web page available only to Boeing employees and order and configure a computer online. Dell suppliers maintain a two-week supply of components close to Dell factories: this inventory belongs to the suppliers, not to Dell. Dell shares information with suppliers on inventory levels, sales and forecasts, and works with suppliers as a virtual enterprise.

Application of B2B to surveying/construction

One of the most discussed topics in e-commerce is business models, because just like conventional business, e-business needs to make a profit and, as many dot.com companies have found to their cost, a key component in this process is a robust business model. Today this is more important than ever as the cold wind of reality hits e-commerce. What is needed now is a proven business case; funders want to see profits before they are prepared to hand out any more money. When a potential investor in an e-commerce project asks the question 'What is your business model?', what they are really trying to discover is: Where is the business going to make money and why are people going to pay for using that particular service? A business model should give product/service, information, income-generation flows and, together with the marketing strategies, enable the commercial viability of the project to be assessed. An e-business model makes it possible to answer such questions as: How is competitive advantage being built and what is the positioning and the marketing mix? In theory, many new business models can be conceived; however, in practice only a limited number are being realised in e-commerce. Figure 4.1 illustrates eleven business models. They are mapped along two dimensions to indicate:

- the degree of innovation, which ranges from (bottom left-hand corner) essentially an electronic way of carrying out traditional business, to value chain integration (top right-hand corner), a process that cannot be done at all in a traditional form, as it is critically dependent on information technology to allow information flow across networks;
- the extent to which functions are integrated.

Other models which have applications in quantity surveying/construction are listed below.

Application Service Providers (ASP)

While the ASP model is new and no settled definition is possible, it may be regarded as a relationship whereby an IT service company, such as Microsoft or Cable and Wireless, manages and delivers applications and/or computer systems to business users remotely, via the internet. This model is to be found in several applications used by quantity surveyors.

E-shop – seeking demand

This can be the web marketing of a company to promote, in the first instance, the services that it provides. Seen as a low-cost route to global presence it is now almost obligatory for quantity surveying practices to have their own website. Its function is primarily promotion, although this has to be sought by prospective clients.

E-mall – industry sector marketplace

An e-mall in its basic form is a collection of websites usually under a common umbrella (e.g. www.propertymall.com). The e-mall operator may not have an interest in the individual sector and income is generated for fees paid by the hosted websites, which usually comprise an initial set-up fee plus an annual fee.

For others the past few years have proved more difficult. In previous editions BuildOnline, an Ireland-based organisation specialising in providing collaboration platforms for construction projects, has been discussed. BuildOnline was one of the early pace-setters in the UK market but it faltered and was soon languishing some distance behind Building Information Workshop (BIW) and 4Projects. In 2007 the company filed a pre-tax loss of £77,000 and the workforce was cut by half; subsequently BuildOnline was taken over by SWORD and renamed FusionLive. The downturn in the construction industry in 2008 also proved difficult for BIW which lost many customers as trading conditions became tougher. Thus instead of open marketplaces it seems likely that the construction industry will gravitate towards more private trading hubs. Increasing consolidation is expected to see each specialist area (collaboration, procurement) dominated by two or three sites.

Perhaps, also, a note of realism has been sounded, as contractors now seem to understand that their core business is not suited to running these kinds of ventures, and as a result they are turning to specialist providers to supply the technology and organisation for them (see discussion below). In addition, there has been an awareness that companies cannot provide every kind of service; consequently, the future for e-commerce seems to lie, at least within the property sector, with niche market provision. One fact that all sectors of e-commerce are sure about is that the more fragmented the market, the more efficiency benefits e-commerce ventures can bring, by uniting the disparate elements of the supply chain.

E-resources; e-procurement

E-tendering

E-tendering enables the traditional process not only to be made more efficient but to add significant value. It can provide a transparent and paperless process, allowing offers to be more easily compared according to specific criteria. More importantly, by using the internet, tendering opportunities become available to a global market.

E-procurement is the use of electronic tools and systems to increase efficiency and reduce costs during each stage of the procurement process. Of all the resources referred to in Figure 4.3, e-procurement is the one that intersects the most with other typologies, often in a complex way. Since autumn 2002 there have been significant developments in e-procurement; legislative changes have encouraged greater use throughout the EU; new techniques such as electronic reverse auctions have been introduced, not (it has to be said) without controversy. In addition, the UK government has launched a drive for greater public sector efficiency following the HM Treasury's publication of the *Gershon Efficiency Review: Releasing Resources to the Frontline* in July 2004, and e-procurement is seen to be at the heart of this initiative.

The stated prime objective of electronic tendering systems is to provide central government, as well as the private sector, with a system and service that replaces the traditional paper-tendering exercise with a web-enabled system that delivers additional functionality and increased benefits to all parties involved with the tendering exercise. The perceived benefits of electronic procurement are as follows:

- Efficient and effective electronic interfaces among suppliers and civil central governments, departments and agencies, leading to cost reductions and time saving on both sides.

- Quick and accurate prequalification and evaluation, which enables automatic rejection of tenders that fail to meet stipulated 'must-have criteria'.
- A reduced paper trail in tendering exercises, saving costs on both sides and improving audit.
- Increased compliance with EU Procurement Directives, and best practice procurement with the introduction of a less fragmented procurement process.
- A clear audit trail, demonstrating integrity.
- The provision of quality assurance information (e.g. the number of tenders issued, response rates and times).
- The opportunity to gain advantage from any future changes to the EU Procurement Directives.
- Quick and accurate evaluation of tenders.
- The opportunity to respond to any questions or points of clarification during the tendering period.
- Reduction in the receipt, recording and distribution of tender submissions.
- Twenty-four-hour access.

Figure 4.6 illustrates the possible applications of e-procurement to projects that are covered by the EU Public Procurement Directives.

Benefits of electronic tendering and procurement of goods and services are said to be wider choice of suppliers leading to lower cost, better quality, improved delivery, and reduced cost of procurement (e.g. tendering specifications are downloaded by suppliers rather than sent by snail mail). Electronic negotiation and contracting and possibly collaborative work in specification

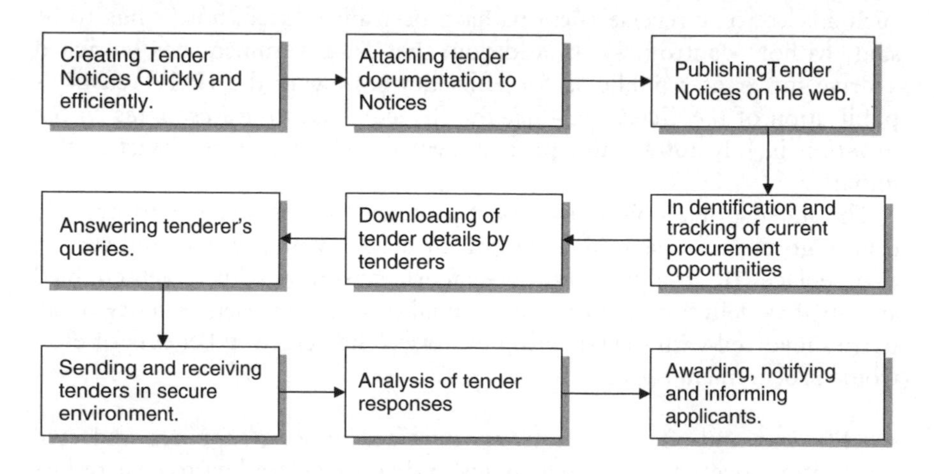

Figure 4.6 Possible applications of e-procurement

can further enhance time and cost saving and convenience. For suppliers the benefits are more tendering opportunities, possibly on a global scale, lower cost of submitting a tender, and possibly tendering in parts which may be better suited to smaller enterprises or collaborative tendering. Lower costs can be achieved through increased efficiency, and in some sectors the time may not be too far distant when the majority of procurement is done this way. However, a survey carried out by e-Business Watch in 2002 over 6,000 organisations found that nearly 60 per cent of those surveyed perceived that face-to-face interaction was a barrier to e-procurement while online security continued to be a major concern.

In autumn 2005 the RICS produced a guidance note on e-tendering in response to the growth in the preparation of tender documents in electronic format. Figure 4.7 sets out their recommended approach to the e-procurement process, while Figure 4.8 maps the way by which contract documentation may be organised for the e-tendering process.

It is worrying that there is still little evidence of the use of electronic

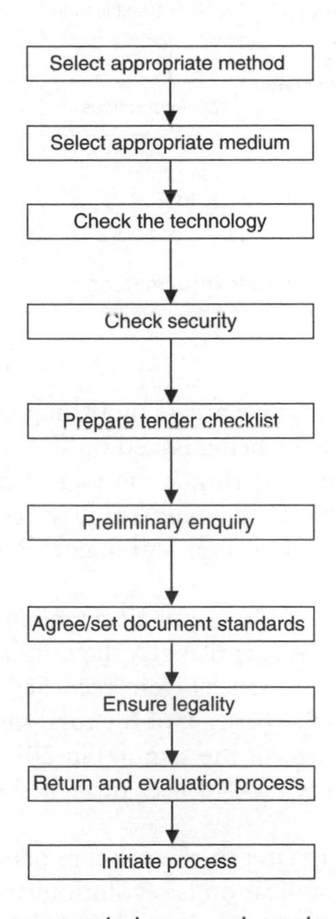

Figure 4.7 The RICS's recommended approach to the e-procurement process

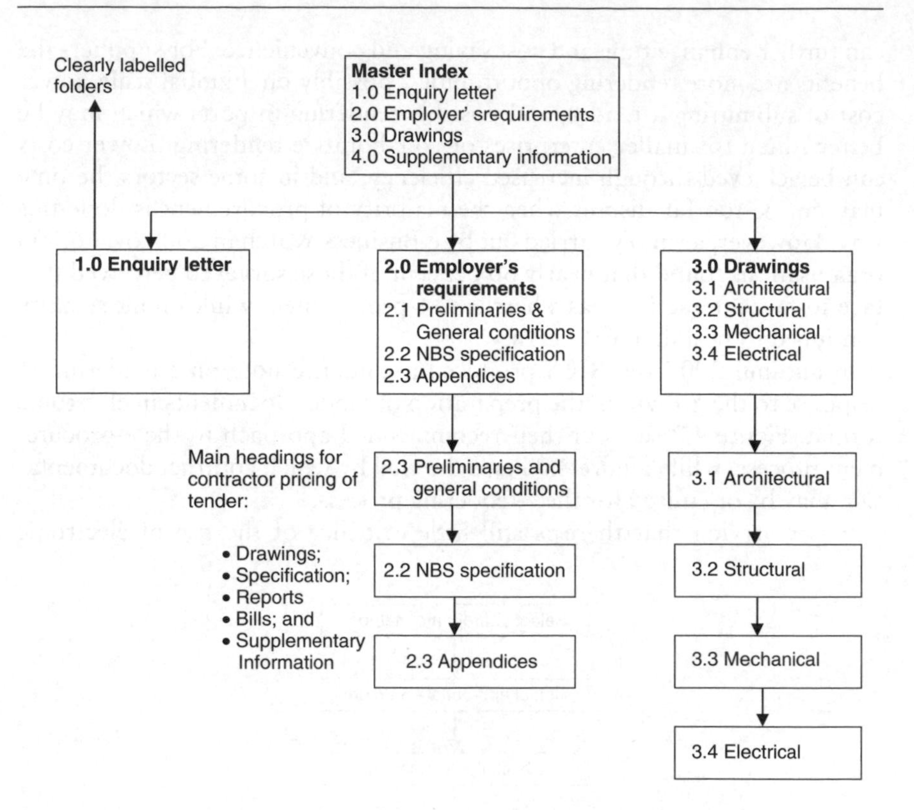

Figure 4.8 Organisation of contract information for e-procurement

Source: RICS.

tendering. It may well be that we are going through a period of paper-based tendering with the documents being issued on CD as well, leading to under-reporting. Extranet-based tendering is the logical next step in the market, using such examples as the RCIS e-tendering service. Future surveys should pick up an increase in usage in e-procurement (RICS Contracts in Use in 2007, 2010).

An additional question on the use of electronic tendering was also asked for the second time. Perhaps surprisingly, the number of instances recorded fell markedly. Only six occurrences were recorded (on projects up to £10 million), representing just 0.4 per cent of the total sample, compared to fifty-four projects (or 2.3 per cent of the sample) in 2004. Under-recording may partly explain the apparent lack of use, though the 2004 sample suffered from a similar deficiency.

Good communication is vital to the procurement and construction process, and electronic communication is revolutionising the means of communication available. However, previous research by the RICS Construction

Faculty has shown that the take-up of electronic communication as part of the procurement process in the UK construction industry has been, at best, patchy. The Construction Faculty's report, *Measurement Based Procurement of Buildings*, showed that while all bills of quantities were prepared in digital form, fewer than 30 per cent were made available to the contractor as an electronic document and fewer than 10 per cent of priced bills were submitted electronically. Furthermore, only 3 per cent of consultants and 4 per cent of contractors had had any experience of e-commerce systems.

As far as construction is concerned, another study carried out by Eadie *et al.* published in 2010 continued to show that the update of e-procurement is still at very low levels in the construction industry. The study indicated that fewer than 25 per cent of construction organisations use e-procurement in the UK which is substantially below other sectors. Can it be that despite the perceived advantages of e-procurement the industry is so conservative and unwilling to change and embrace new approaches and techniques?

The research went on to try to define the drivers and barriers to the introduction of e-procurement in construction and understand why this is the case. One reason often cited is the uniqueness of the construction industry compared with manufacturing; perhaps there is some element of truth in this, but it cannot completely account for the different levels of uptake. A review of current practice in 2010 identified that the perceived drivers and barriers to the use of e-procurement may be categorised as shown in Table 4.1.

Of 483 surveying organisations only eighty-three were using e-procurement, although this figure covers a wide disparity between the public and

Table 4.1 E-procurement drivers

E-procurement drivers	*Category*
Cost	Cost savings in administration and transactions. Increased profit margins. Strategic cost savings.
Time	Shortened procurement times. Shortened communication times. Shortened times through greater transparency. Reduction in evaluation times. Shortened contract completion time.
Quality	Increased quality through benchmarking. Increased quality through the supply chain. Increased efficiency. Increased quality through improved communications. Increased accuracy.
General	Gain competitive advantage. Convenience of archiving completed work. Develops technical skills and expertise of procurement staff.

private sectors, with some 74 per cent of organisation within the public sector engaging with e-procurement, while in the private sector fewer than 25 per cent of organisations were using e-procurement. The rank order of drivers and barriers varies between the public and private sector, although 'Cost savings in administration and transactions' and 'Security in the process' were ranked number one by both sectors (Table 4.2). Overcoming the most important barriers and incorporating the most important drivers within e-procurement systems will achieve a higher level of maturity.

E-auctions

An online auction is an internet-based activity which is used to negotiate prices for buying or selling direct materials, capital or services (Figure 4.9). Online auctions that are used to sell: these products are called forward (or seller) auctions and closely resemble the activity on websites such as Ebay;

Table 4.2 E-procurement barriers

E-procurement barriers	*Category*
Cultural	Lack of management support/leadership. Resistance to change. Lack of widely accepted e-procurement software solution. Magnitude of change. Lack of national IT policy relating to e-procurement issues. Bureaucratic dysfunctionalities. Lack of technical expertise. Staff turnover.
Infrastructure	Access to internet. Insufficient assessment of systems prior to installation.
Security	Security in the process. Confidentially of information – unauthorised access. Data transmission reassembly – incorrect reassembly of transmitted data. Incomplete documents supplied.
Legal	Lack of pertinent case law. Different national approaches to e-procurement. Proof of intent – electronic signatures. Clarity of sender and tenderer information. Enforceability of electronic contracts.
Assessment costs	Information technology investment costs.
Compatibility	Internal and external interoperability of e-procurement software. Investment in compatible systems. Reluctance to 'buy-into' one off systems.
General	Perception of no business benefits realised. Lack of awareness of best practice solutions. Lack of forum to exchange ideas.

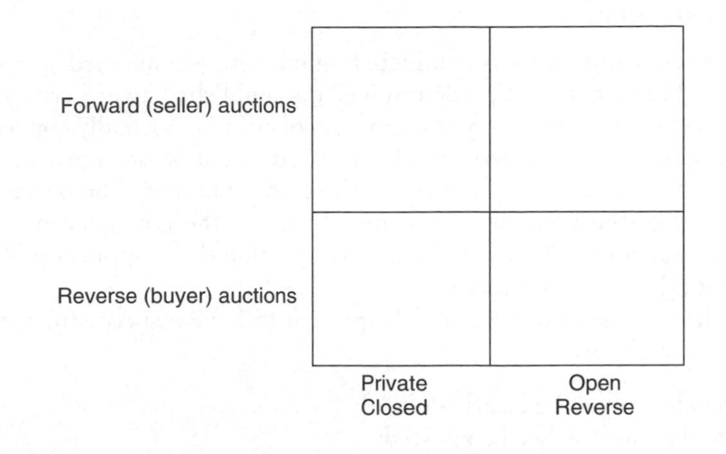

Figure 4.9 An online auction

the highest bidder wins. Some companies are starting to use reverse auctions where purchasers seek market pricing, inviting suppliers to compete for business on an online event. Auctions may either be private/closed where there are typically few bidders who have no knowledge of each other's bids, or open where a greater number of participants are invited. In this case participants have prior knowledge of either their rank or the bidding itself. When used, the technique can replace the conventional methods of calling for sealed paper tenders or face-to-face negotiations.

Online auctions are said to offer an electronic implementation of the bidding mechanism used by traditional auctions and systems may incorporate integration of the bidding process together with contracting and payment. The sources of income for the auction provider are from selling the technology platform, transaction fees and advertising. Benefits for suppliers and buyers are increased efficiency and time saving, no need for physical transport until the deal has been established, as well as global sourcing.

There have been some strong objections to e-auctions and in particular reverse e-auctions from many sections of the construction industry. In the public sector the OGC has received representations from trade associations and other bodies. Sections of the industry have seen e-auctions as a return to lowest price purchasing, threatening already low margins. The industry also perceives e-auctions as challenging the principles of many government-led initiatives (see best practice portals) such as an integrated supply chain approach to construction procurement based upon optimum whole-life value. Among quantity surveyors the perceptions of era are very negative, with 90 per cent believing that they reduce quality and adversely affect partnering relationships.

Reverse e-auctions

The reverse e-auction event is conducted online with prequalified suppliers being invited to compete on predetermined and published award criteria. A reverse e-auction may be on any combination of criteria, normally converted to a 'price equivalent'. Bidders are able to introduce new or improved values to their bids in a visible and competitive environment. The procedure and duration of the event will be defined prior to the commencement of the reverse e-auction. There will be a starting value that suppliers will bid against until the competition closes.

Three characteristics that need to be present to have a successful reverse auction are as follows:

- the purchase must be clearly defined;
- the market must be well contested;
- the existing supply base must be well known.

These three factors are interdependent and together form the basis for an auction that delivers final prices as close as possible to the true current market price. For both the buyer and the supplier a clearly defined scope of work is essential; without this it becomes very difficult accurately to bid for the work. Contestation (that is, three or more suppliers within the market willing to bid for the work) is another prerequisite. Without this there isn't any incentive for suppliers to reconsider their proposals. Finally the client's knowledge of the supply base ensures that the most suitable suppliers participate in the events. In order to move away from a system where cost is the only selection criteria it is possible to organise bidders to submit, before the commencement of the reverse auction, their proposals on other matters, such as safety or technical ability. These proposals can then be evaluated beforehand and the resulting scores built into the auction tool. Therefore, when a supplier enters their price, the application already has the information needed to complete the evaluation process. This process is known as a transformation reverse auction (TRA) and is believed to more accurately reflect the prevailing market dynamics.

E-projects

- Project extranets
- Supply chain management.

E-commerce may be said to be the ultimate supply chain communication tool, as it permits real-time communication among members. e-commerce may include the use of some or all of the following technologies:

- ***The internet*** – the international network. The main advantages of the internet for the quantity surveyor include availability, low cost and easy

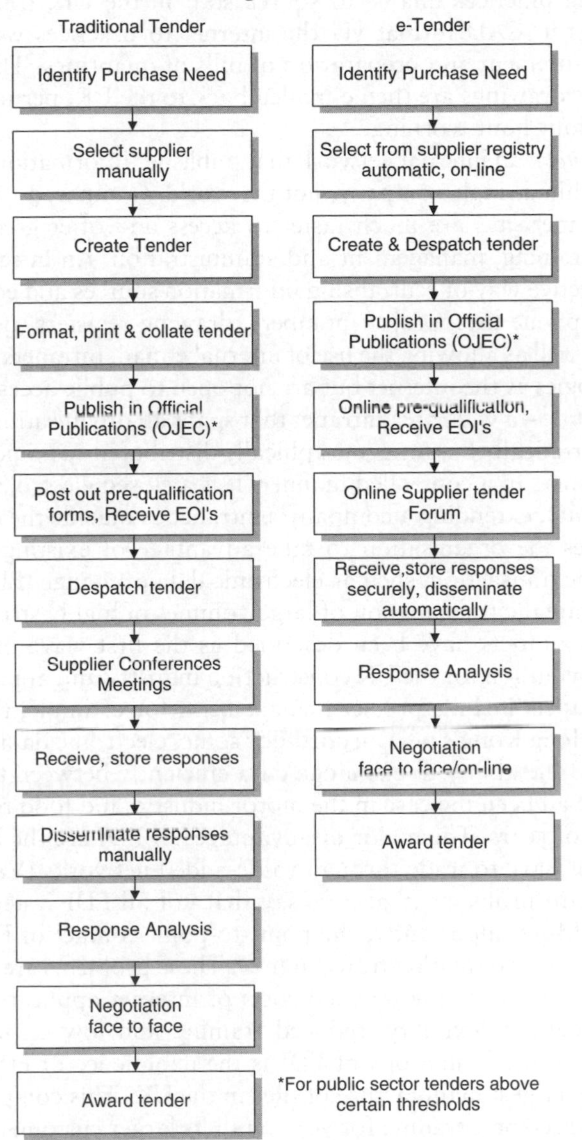

Figure 4.10 E-tendering versus traditional tendering

access, whereas the main disadvantages, particularly for business users, centre around lack of control, reliability and security, aspects that are now being addressed and will be discussed later in this chapter. For the surveyor, internet applications include procurement, marketing, e-mail and data transfer. A significant number of hard-pressed UK quantity surveying practices unable to source staff in the UK, transfer project drawings in CAD format via the internet to practices worldwide for the measurement and preparation of bills of quantities. The completed bills plus drawings are then e-mailed back to the UK, permitting virtual twenty-four-hour working.

- *An intranet* – an internal network that publishes information available to staff within a single company, not the world. Compared with the internet, intranet sites are much faster to access and offer great savings in set-up, training, management and administration. An intranet is a very cost-effective way of centralising information sources and company data such as phone lists, project numbers, drawing registers, quality procedures as well as allowing the use of internal e-mail. Intranets use the same technologies as the internet but are not open to public access.
- *An extranet* – a wide area intranet that spans an organisation's boundaries, electronically linking geographically distributed customers, suppliers and partners in a controlled manner. It is a closed electronic commerce community, extending a company's intranet to outside the corporation. It enables the organisation to take advantage of existing methods of electronic transaction, such as electronic data exchange (EDI), a system to facilitate the transmission of large volumes of highly structured data. Project extranets have been described as the first wave of the e-commerce revolution for the UK construction industry and applications that use extranets include project management, for example, the construction of Hong Kong's new airport. For some, electronic data interchange promised the ability to exchange data efficiently between trading partners, as had been the case in the motor industry and food retailers for a number of years. The major disadvantages of EDI are the high cost, as operators have to trade through value-added networks (VANs) as well as standard problems; that is to say that not all EDI systems are compatible. More importantly, the point-to-point contact of EDI provides no community of market transparency. These problems are increasingly being addressed by the reduced costs of internet applications that are able to deliver flexibility, reduced training and low set-up costs. An example of the limitations of EDI is the experience of H&R Johnson Tiles, the largest manufacture of tiles in the UK. This company has carried out electronic trading for years with its larger customers using EDI; however, the company has found it impossible to extend EDI to smaller customers without the critical mass of transactions to drive the necessary investment; a separate website had to be launched using an extranet to cater for its top-twenty smaller customers. This is not to say that batch-mode EDI transactions will not survive and prosper, as the system is extremely efficient and has been predicted to have an established place in the large-scale exchange of data.

Extranets/EDI are perhaps the most diffused type of e-commerce and should be better defined as e-business rather than as e-commerce.

Collaboration platforms

These provide a set of tools and information environment for collaboration between organisations. This can focus on specific functions, such as collaborative design and engineering or provide project support with a virtual team of consultants. Business opportunities are in managing the platform and selling the specialist tools.

Porter's value chain

The value chain (see Figure 4.11) has been defined as a model that describes a sequence of value-adding activities of a single organisation, connecting an organisation's supply side with its demand side, and includes supporting activities. Information technology is applicable at all points of the value chain.

Primary activities include:

- inbound logistics, just in time activities;
- operations, process control;
- outbound logistics, online link to customers;
- marketing and sales, laptops for direct sales;
- service, electronic dispatch of technical supports.

Support activities include:

- firm infrastructure, e-mail;
- human resource management, online personnel base;

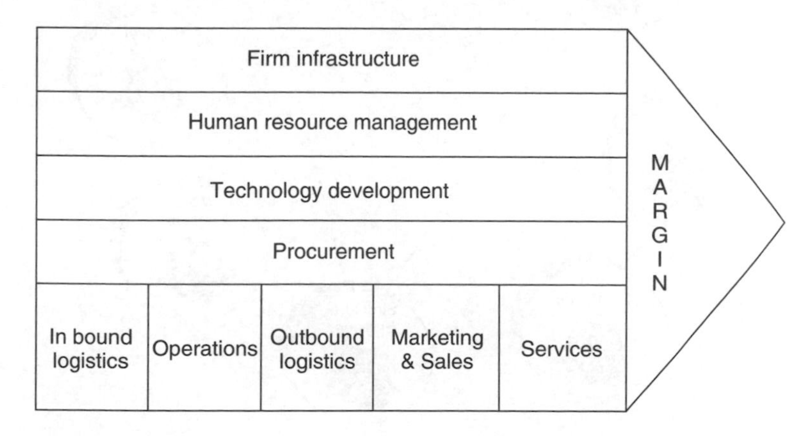

Figure 4.11 The value chain

- technological development, CAD/CAM;
- procurement, online access to suppliers' inventory.

Value chain integration can use internet technology to improve communication and collaboration between all parties within a supply chain.

Value chain service provider

These organisations specialise in a particular and specific function of the value chain; for example, electronic payment or logistics. A fee or percentage-based scheme is the basis for revenue generation.

Value chain integrators

These focus on integrating multiple steps in the value chain, with the potential to exploit the information flow between these steps as further added value.

Virtual communities

Perhaps the most famous of virtual communities is the ubiquitous Amazon.com. Figure 4.12 illustrates the traditional lines of communication

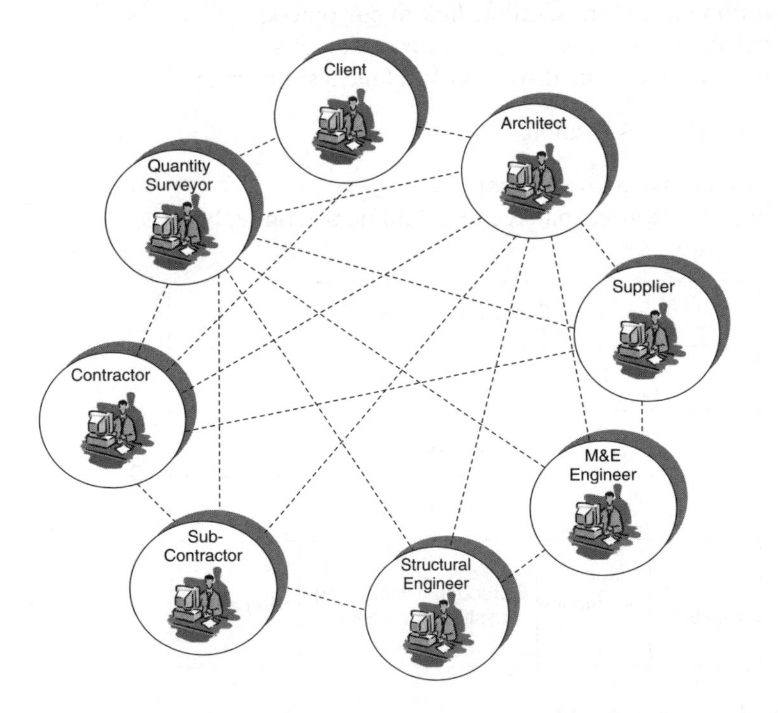

Figure 4.12 Traditional commercial networks

between the various supply chain members. Without the presence of a hub to link and coordinate the various parties, decisions can and frequently are taken in isolation, without regard for the knock-on effects of cost, delays to the programme, impact on other suppliers and so on.

Traditional communication networks

However, by contrast, the introduction of a collaborative hub (for example, the packages offered by SWORD) permits decision-making to be taken in the full light of knowledge about the possible implications of proposed changes. In addition it also permits specialist subcontractors or suppliers to contribute their expertise to the design and management process.

Collaboration platform communication

In a recent case study the quantity surveying practice Gleeds calculated that the drawing production cost savings arising from the use of an electronic data management collaboration system used on a £5 million, thirty-week retail construction project were as follows:

Printing cost for project drawings	£46,112
Postage for drawings	£1,584
Copying costs for project specification	£10,215
Postage costs for specification	£219
Total:	£58,130

Source: CITE.

Knowledge management

'A learning organisation is an organisation skilled at creating, acquiring, and transferring knowledge, and at modifying its behaviour to reflect new knowledge and insights' (Garvin, 1998).

Knowledge management is a comparatively new concept, becoming established in the early 1990s, and may be defined as a means by which to capture and monitor the ever-increasing bodies of intellectual knowledge, and facilitates the efficient creation and exchange of knowledge on an organization-wide level. Knowledge management should not be confused with information management. Information management consists of predetermined responses to anticipated stimuli. Knowledge management, however, consists of innovated responses to new opportunities and challenges. Knowledge management typically focuses on organisational objectives such as improved performance, competitive advantage, innovation, the sharing of lessons learned, integration and continuous improvement. Knowledge management can help

individuals and organisations share practices and approaches in order to reduce redundant work and avoid reinventing the wheel. It can also reduce training time for new employees, retain expertise when employees leave an organisation, and help adapt to changing environments and markets.

Sources of knowledge

During the course of a day a quantity surveyor will draw upon many sources of knowledge when, for example, providing cost and contractual advice for clients or preparing procurement documentation. The larger the organisation the larger the potential store of knowledge and the more important it is that the knowledge available is effectively managed in order to add value to the organisation as well as the service that clients receive. These knowledge sources may be categorised as follows:

- Explicit: knowledge that is easily codified and conveyed to others. It is articulated knowledge, expressed and recorded as words, numbers, codes, mathematical and scientific formulae. Explicit knowledge is easy to communicate, store and distribute, and is the knowledge found in books, on the web, and other visual and oral means, for example; standard forms of contract and methods of measurement. It is the opposite of tacit knowledge. It is occasionally referred to as hard knowledge.
- Tacit: the concept of tacit knowledge was introduced by the Hungarian philosopher-chemist Michael Polanyi (1891–1976) in his 1966 book *The Tacit Dimension*. By comparison to explicit knowledge, tacit knowledge is difficult to formalise and convey to others. Tacit knowledge is often based upon experience, instinct and personal insights; for example, recognising the tell-tale signs of financial difficulties within a contractor's or a subcontractor's organisation that may develop into bankruptcy or insolvency. It is occasionally referred to as soft knowledge.
- Implicit: this form of knowledge falls between explicit and tacit and is knowledge that was once explicit but over time is capable of being codified and quantifiable. It may mean taking a second look at tacit knowledge to determine whether it can be codified.

According to Davis *et al.*, the quantity surveying profession is characterised by a wealth of experiential knowledge which is tacit and cannot be written down easily. It is crucial that quantity surveying firms realise their true potential assets, which can be determined by the introduction of a knowledge management system as it enables the company to 'know what it knows'. Quantity surveyors are professionals who provide help to clients for the legal and financial problems with their expertise. The more projects quantity surveyors complete the more experience they gain. However, most QS firms face a problem in that they are losing knowledge due to the retirement or resignation of key personnel. With the help of a knowledge management system, knowledge is shared and stored, and thus the risk of losing the knowledge can be minimised.

How can knowledge management be introduced into a quantity surveying practice?

A prerequisite on introducing a knowledge management system into an organisation is to develop a culture where employees are willing to share their knowledge and not be victims of a blame culture. Like many new approaches to management there will be an unwillingness within an organisation to change current operating practices and evaluate the perceived benefits.

Knowledge management processes and the potential for IT

The ways in which knowledge management can be introduced will depend on the size and type of an organisation. BSI PD7503 – Introduction to Knowledge Management in Construction identifies seven crucial aspects:

- Decide what is required from the knowledge management programme.
- Draw up a strategy for implementation.
- Understand the organisation's current knowledge resources.
- Enable a sharing culture.
- Manage the knowledge content.
- Use enabling technology.
- Measure and evaluate the results.

Table 4.3 indicates a number of knowledge management processes and the potential for IT. The knowledge management processes may be said to be:

- Knowledge creation.
- Knowledge storage/retrieval.
- Knowledge transfer.
- Knowledge application.

In a survey of quantity surveyors in 2007 Davis *el al.* approached fifty quantity surveyors in practice to discover their perception of knowledge management. The majority of quantity surveyors questioned agreed that knowledge management would bring numerous benefits to the company and to themselves, and were very positive. However, they were not convinced that an increase in innovation would occur. This may be explained by the lack of emphasis on creating knowledge. Furthermore, quantity surveyors expected that knowledge management would bring personal benefits more than

Table 4.3 Knowledge management processes and the potential for IT

Knowledge management processes	*Knowledge creation*	*Knowledge storage/ retrieval*	*Knowledge transfer*	*Knowledge application*
Supporting information Technologies	Data mining Learning tools	Electronic bulletin boards Knowledge repositories Databases	Electronic bulletin boards Discussion forums Knowledge directories	Expert systems Workflow systems
IT enables	Combining new sources of knowledge Just-in-time learning	Support of individual and organisational memory Inter-group knowledge access	More extensive internal network More communication channels available Faster access to knowledge sources	Knowledge can be applied in many locations More rapid application of new knowledge through workflow automation
Platform technologies	Intranets/groupware and communication technologies			

benefits to the company. This is shown by the current knowledge-sharing practices they frequently used. The findings illustrate that in the knowledge-sharing process, knowledge and experience are mainly personal and not company tactical experiences, or company-level problem-solving techniques.

The majority of the respondents believe that the most critical factors to knowledge management success are management support, employee active participation, application of IT systems and creating knowledge-sharing space.

Again, the findings indicated that the main barriers for a quantity surveying firm to develop a knowledge management system are lack of time and understanding of the processes involved, and the difficulty to locate, capture, generalise and store knowledge. The resistance to change the current practice and employ new management approaches from top management also contributes to the opposition to develop knowledge management in firms. As previously stated, enterprises enter supply chains to improve profitability through product development that is responsive to consumer demands. Increasing market share and maintaining a competitive product depends on the ability of each unit in the chain to apply knowledge innovatively. The key to innovation is the application of explicit and implicit knowledge of the people within the organisations throughout the chain. With improving technology and a supportive learning environment, enterprises in the supply chain can add value by using the collective wisdom of the people in the organisation.

The quantity surveyor and e-commerce

For the purposes of this review e-business has been broken down into the following:

1. How the surveyor can participate in e-commerce.
2. Guidelines for the integration and adoption of e-commerce practice.

The DETR report *IT Usage in the Construction Team* found that although the majority of information in the construction industry is created using IT, most is distributed in paper form. The report found that:

- Seventy-nine per cent of specifications are produced electronically against which 91 per cent are distributed on paper.
- Seventy-three per cent of general communications across the design team are produced electronically, yet 85 per cent are distributed on paper.
- Only 5 per cent of contractors' tenders are received in electronic format.

It is little wonder that the European construction industry has been estimated to spend £300 million a year alone on couriers! In addition, SWORD claims that lost paperwork and lack of communication adds 20 to 30 per cent to construction costs across the board. Tables 4.4 to 4.7 have been included in order to give the surveyor some indication of the potential for the integration of e-commerce into day-to-day practice. Three levels have been proposed based upon commitment, ranging from level one, requiring minimal investment, but nevertheless still at a level capable of producing tangible benefits and savings to both clients and business, through to level three, the virtual organisation, requiring total commitment and a high level of investment.

Levels of e-business

Entering the world of e-commerce should be a considered business decision, not a knee jerk reaction to the sudden availability of new technology. An existing bricks-and-mortar company's goals are different from those of an ideas company. Likely to be already profitable, a bricks-and-mortar company will most often turn to the internet to expand its markets, meet customers' needs and improve operating efficiencies in ways that are usually incidental to an existing business – not to reinvent its business. The first step is to identify the business goals that may be served by using the internet. Answering this question depends in part on the nature of the business. For example: How large is the company? What sort of distribution chain does it use? Is its customer base static or could it be expanded? Some examples of

Table 4.4 Levels of e-business

Level	*1*	2	3
Capability	Electronic mail and World Wide Web	Information management (Electronic Procurement)	Virtual Organisation (e.g. Dell, Cisco, Amazon)
What is needed?	Personal computer Internet access Software	Process redesign	Business redesign Culture change
Cost	Minimal investment Cost of running parallel systems	Change management	Change management
Benefit	Easy access to information Less information processing	Competitive advantage Streamlined business processes Alignment with supply chain partners	International competitiveness Best practice operations Robust relationships
Who benefits?	SMEs	Public sector Industry	Public sector Industry All members of the supply chain

potential e-commerce applications to surveying practices over levels one to three are given in Tables 4.5 to 4.7.

Even at the entry level considerable market advantage may be gained through the use of e-mail and interactive websites.

Table 4.5 E-business – Level 1

Applications	*Benefits*
1. E-mail	1. Real time communication
2. Marketing footing to larger organisations	2. Access to new markets on similar
3. Participation in electronic procurement / auctions	3. Clients may increasingly insist on electronic procurement
4. Access to databases; for example:	4. Information on competitors/market opportunities
TED Tenders Direct Barbourexpert Constructionline	Sourcing using web pages

Table 4.6 E-business – Level 2

Applications	*Benefits*
1. *Fully electronic procurement including*: Pre-qualification. Comparing bids. Evaluation. Contract award. Archive information on bidders including KPI rating. Online auctions	1. By far the biggest cost savings come from the reduced costs of creating and disseminating tenders; evaluation of bid responses, creating purchase orders and tracking progress. Access to benchmarked records on contractors and supply chain.
2. *Fully electronic tendering including:* Tender submissions. Submission of bids. Online estimating. Exchange of Information. Payment.	2. Reduce wastage, increase profit margins. Reduce errors.
3. *Project management* Dissemination of information.	3. Effective supply chain management. Real-time communication. Integration of the supply chain.
4. *Enables virtual design teams* Collaboration with practices worldwide.	4. 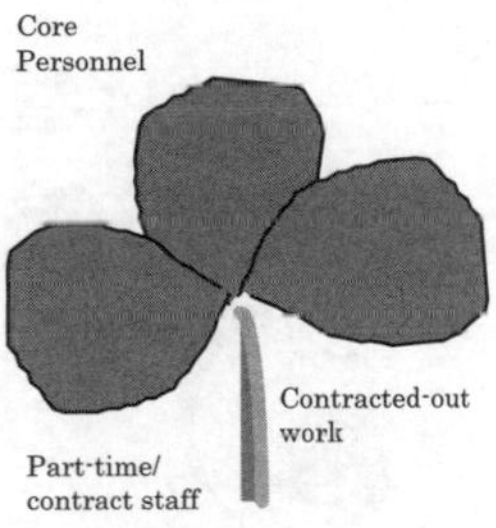 Core Personnel Contracted-out work Part-time/ contract staff Contracted-out staff could be located anywhere on the globe.

The choice of partner can be crucial in maximising opportunities; for example, by working with existing operators it is possible to start benefiting immediately.

A constant theme throughout this book has been client criticisms of the UK construction industry and its fragmented structure. Clients and professionals as well as the entire downstream supply chain try to cope with the challenge of operating in a highly fragmented industry, where the top five contractors own less than 10 per cent of the marketplace. The fragmentation has in return over the years contributed to poor profitability and cashflow,

Table 4.7 E-business – Level 3

Applications	*Benefits*
1. E-construction packages including: Project collaboration packages allowing construction professionals to access and amend project information.	1. E-construction packages: Alter and amend project documentation at minimal cost. Ensures that everyone works on the most up-to-date information. A repository for comments. Provides accurate audit trail. Compresses the construction programme. Few change orders.
2. Industry and supply side marketplaces.	2. Real time information of stock levels/delivery, etc.
3. Contractor consortia.	3. Pooling and sharing of information.
4. Niche markets.	4. Dispute resolution.

even for the major players, which have in turn prevented investment in new technologies as noted by Latham and Egan. Add to this the constant demand by clients for added value and the case for adopting e-commerce solutions and practices seems irresistible. As demonstrated in Figure 4.13 the biggest

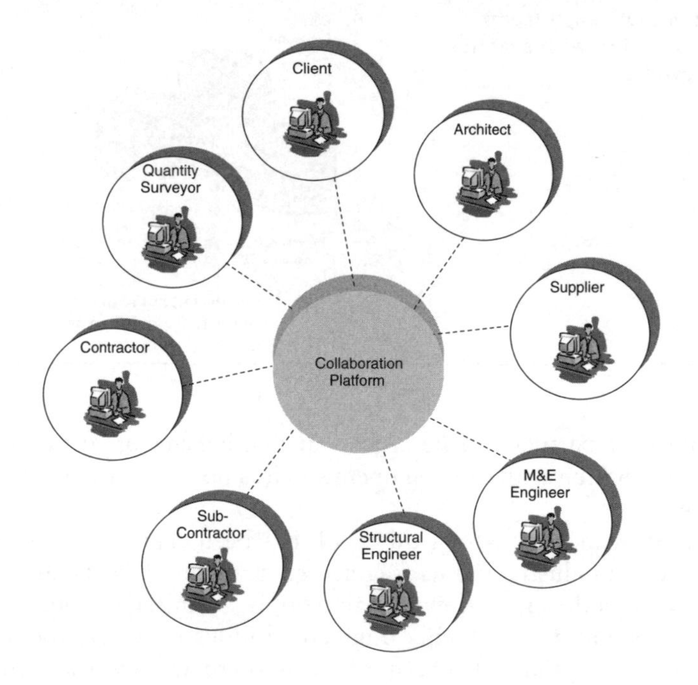

Figure 4.13 Collaboration platform

benefits of e-commerce are likely to come from the integrated supply chain, where information is freely available among clients, contractors and suppliers.

Electronic document management systems (EDMS)

Of the many systems that emerged from the early 2000s, some failed to capture the imagination and fell by the wayside, others were clearly ill-conceived, while others still survive and try to convince the profession of their usefulness.

On the face of it electronic data management would appear to be the ideal solution for the quantity surveyor who on a daily basis has to deal with a large volume of information in a variety of formats. While quantity surveyors can never be regarded as Luddites when it comes to embracing IT solutions and intregating them into traditional situations, there appear to be limits as to how far IT can penetrate the quantity surveyors' working practices.

EDMS vary in complexity and functionality. The nature of the construction industry presents a number of problems relating to the electronic management of documents; these, according to the RICS, may be said to be:

- Because information is primarily centred around individual projects, rather than in set business structures. This increases the management and control required.
- There are large information flows both in and out of organisations to many different external organisations and these flows vary from project to project.
- Drawings and models pose particular problems due to the size of files involved.
- Documents that have been executed under seal may need to be retained for twelve years.
- There are numerous legal requirements on organisations that hold and control information, in particular the Data Protection Act 1998 and the Freedom of Information Act 2000, although this legislation is not exclusive to data stored and managed electronically.

The typical order for the process illustrated in Figure 4.14 is to:

- create
- capture
- share
- collaborate
- control
- index

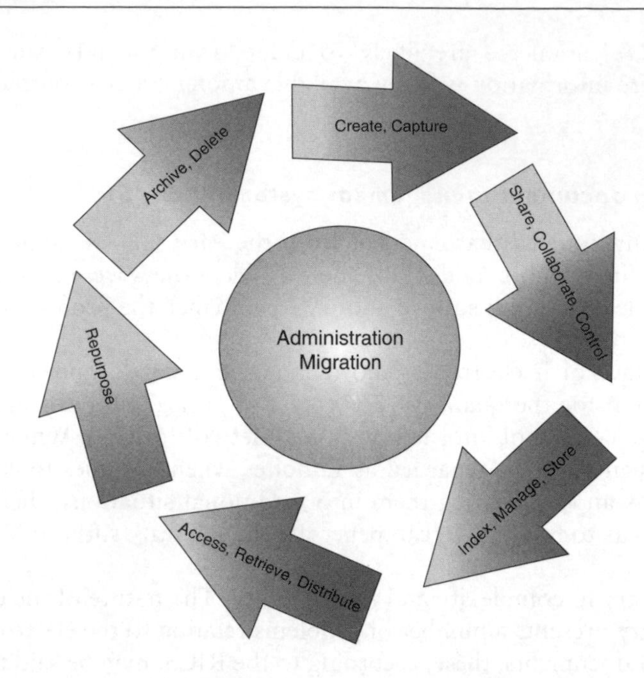

Figure 4.14 Document life cycle; RICS Practice Note Electronic Document Management

- manage/store
- access
- retrieve
- distribute
- repurpose
- achieve
- delete.

It would seem to be that the key to implementing an efficient EDMS is rigorous evaluation according to the level of sophistication and functionality.

Legal aspects of e-commerce

Beyond the basic business considerations a host of legal issues face an industry or profession preparing to go online. This review of the implications of e-commerce for surveying practice cannot be concluded without mentioning some of the practical legal considerations that must be taken into account.

One aspect of e-trading has caused considerable concern in both B2B and B2C sectors, namely the regulation of transactions in a global market, where organisations as well as governments can find themselves dangerously exposed in what has been described as a virtual 'Wild West' environment. There is also the view, held by many in the IT sector, that the infrastructure is not robust or sufficiently mature enough to do the job being asked of it as a result of the e-commerce hype.

The intersection of a global medium, like the internet, with systems and legislations designed for the physical, territorial world, poses many problems. Compared with other entities the internet has developed in a spontaneous and deregulated manner and does not have a central point of authority. In the event of contested claims and possible litigation the fundamental problem of jurisdiction remains unresolved, as does the security of the systems, although as previously mentioned steps are being taken to establish the integrity of cyberspace with the introduction of cryptography and digital signature services. Its technical development has been guided by protocols established through bodies such as the Internet Engineering Task Force, but there has not been a central rule-making body that has exercised comprehensive legislative authority over the internet, and there is unlikely to be one. The multi-jurisdictional and multi-functional nature of the internet means that, inevitably, many different interests in many different parts of the world will be concerned with any endeavour to formulate specific policies. Even in the European Union, the suggestion contained in the Commission's directive that e-commerce should be governed by the law of the country where the service provider is established has been questioned by consumer groups that want the local law where the website is accessed to be given priority.

Another crucial area of concern, primarily for governments, is e-tax or the tax treatment of online transactions. At present the volume of e-transactions means that the fiscal implications are modest. However, if predictions of growth are to be believed the question of how and which government is to tax such revenues could be very important. Within the OECD area views diverge; in the United States the belief is that e-transactions should not be taxed, while in the European Union the view is that VAT should be levied on e-trading. The law in the field of e-commerce is continuously developing and fast moving, as numerous drafts pass into the statute books. The evolution of technology also means that legislation must be updated and requires constant review. Organisations need to be aware of both the current and the prospective impact of legal provisions.

The principal regulatory concerns are focused in four areas: online contracting and security, both dealt with previously in the review of encryption and private key services, as well as regulation/jurisdiction and intellectual property protection.

The implementation of electronic keys was dealt with in an EC Framework Directive implemented after July 2001. The UK E-Communications Act 2000, when originally drafted, contained powers to be vested in law-enforcement agencies, requiring disclosure of electronic keys to decrypt information where necessary. This provision caused considerable debate, as it was seen by many as a major barrier to the promotion of the UK as a favourable environment for e-commerce. It was suggested that the Home Office had hijacked the Bill and eventually the government was forced to delete the provisions, although in practice they were only moved to the Regulation of Investigatory Powers Bill, which is now on the statute books. The UK e-Communications Act is therefore quite simple, in that it allows for:

- The introduction of a new approvals regime for providers of cryptography services.
- Electronic signatures to be admissible in court.
- The updating of existing legislation to allow the use of electronic communications.

In the early days of e-commerce it is true to say that the legislative framework was lagging behind business practice. How can an organisation be sure that information being transmitted electronically is secure from its competitors?

The major legislative instruments in e-commerce law are:

- The UNICITRAL Model Law – 1966
- The UK Electronic Communications Act – 2000
- The Electronic Commerce Directive (00/31/EC) – 2002.

It does however only apply to services supplied by services providers within the EU. Countries outside the EU are covered by UNCITRAL rules. The e-commerce Directive establishes rules in the areas including definition of where operators are established; transparency obligations for operators; transparency requirements for commercial communications; conclusion and validity of electronic contracts; liability of internet intermediaries; and online dispute settlement. Put simply, the Directive states that service providers are subject to the law of the member state in which they are established or where the ISP has its 'centre of activities'. B2B contracts are of particular importance as national laws govern the main aspects of contract law, and what constitutes an offer or an acceptance varies from country to country. For example, in the rules formed under English law on offer and acceptance, at what point is the offeree's acceptance communicated to the offerer? Communication by website is instantaneous, while e-mail is not; the possible scenarios include:

- The offerer fails to collect e-mail from the server.
- Failure of ISP.

European Directive 00/31/EC attempts to clarify the situation by stating that the contract is concluded when the offeree is able to access the offerer's receipt of delivery. Unfortunately this clause does not cover the position, say, of an invitation to tender which is an invitation to treat.

In general, the parties to the contract should agree by what is called private autonomy as to which countries the law of contract is to apply.

Contrary to popular belief, there is no legal reason to prevent a bidding legal contract from being made by way of e-mail provided that there is a clear offer and acceptance of all the critical terms. Therefore, within a quantity surveying practice, for example, employees should be reminded of any limits on their authority to conclude contracts online, and this could extend to adding a standard disclaimer to e-mails to this effect. Recent legislation in the B2C sector implemented by the UK government now applies to contracts that are concluded at a distance between, say, a construction supplier and a consumer. The website of the supplier must now include the following:

- the identity and postal address;
- characteristics of the relevant goods/services;
- the price, delivery costs and taxes;
- the existence of the right of withdrawal from the contract.

Failure to comply with these conditions gives the consumer the right to cancel the contract within seven days.

Conclusion

There can be no doubt that the future for both business in general and for the quantity surveyor in particular is as part of the digital economy. In the late 1990s revolutionary new business models were set to destroy old economic values but only two years later the talk of the collapse of the new economy was just as overstated. However, to equate the downturn in the e-economy with the demise of the internet is, like the pundits who proclaim the death of the quantity surveyor, very exaggerated! The market is maturing and, as has been demonstrated in this chapter, the shift towards the digital economy is unstoppable and there are real benefits that can be brought to a fragmented industry by e-commerce.

Therefore despite all of the pitfalls, both technological and legal that have been outlined in this chapter, what are the critical success factors to be considered by an organisation still determined to be part of the digital revolution? First, some reassurance; it would also appear that e-commerce security concerns may have been overstated. As indicated in Figure 4.15 latest data indicate that the most widespread information society breaches are computer virus infections. Almost all organisations have been affected by computer viruses; by comparison the numbers of businesses affected by other security breaches such as unauthorised access to their networks or identity theft are fairly low.

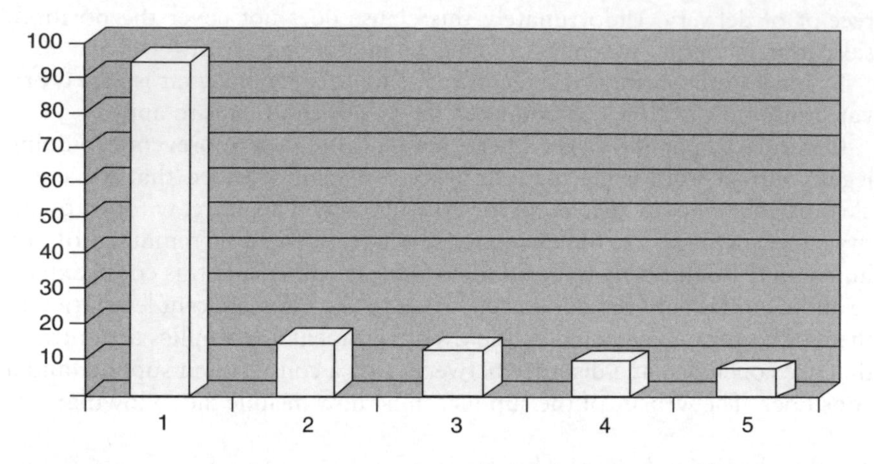

Figure 4.15 Types of security breaches, expressed as a percentage

Key:
1. Computer virus infections
2. Unauthorised entry to internal networks
3. Manipulation of software applications
4. Identity theft
5. Online fraud

Source: SIBIS.

Therefore, assuming that the starting point for a new e-commerce venture is all-round e-commerce (see Table 4.5), the points to consider are as follows; these will vary in line with the level of entry.

Content – This should be a unique service that exploits the electronic environment and delivers added value to potential users. The presence of a unique and/or innovative product or service that is saleable over the internet; that is, it fits with the demographics of the internet. In addition, the platform must be capable of attracting sufficient clients to generate cashflow to repay start-up and running costs (for example, dispute resolution online).

Community – The ability to build up a critical mass of customers/business partners for the venture which will translate into sales/cost savings to cover the initial investment. It is the inability to establish a community of clients that has caused so many dot.coms to fail.

Commitment – Clear objectives. Most clearly demonstrated by a defined business case for the e-commerce venture, but at the very least a clear idea of objectives and the demonstration of strong motivation for using the internet. For example, just-in-time production for building materials – a large volume of curtain walling could be produced only when it is required for incorporation into a project. The manufacturer in return can tap into a global supply base for raw materials.

Control – The extent to which e-commerce is integrated with the internal business process, enabling the organisation to control all aspects of its business and handle growth and innovation.

Website design

The address of the website is crucial as it will influence the number of potential clients who are able to find it. Although domain name registration is relatively cheap and quick, about £13 per annum for a '.co.uk' and slightly more for a '.com' address, time and money can be lost if a similar name has already been registered by a third party. The English courts have given firm judgments against cyber-squatting which is the practice whereby a person with no connection to a specific domain registers it, usually with the intension of selling it on to the highest bidder who has a connection with it. However, except in the case of cyber-squatting the law is still very much 'first come, first served'. When using a third party for website design it is important to establish a formal contract setting out the obligations of designer and client. In particular the ownership of the intellectual property rights should be established, which under normal circumstances should rest with the client. Website content must comply with the Data Protection Act. Five years ago there were five or six major search engines; now the market is dominated by one – Google. The message is clear: a site will not become noticed if it is not ranked by Google. Five or six years ago submission of a website to the search engines was free of charge; today most search engines require upfront payments of several hundreds of pounds to guarantee consideration and ranking.

Bibliography

BSI (2003). *PD7503 – Introduction to Knowledge Management in Construction*. British Standard Institution.

Building Centre Trust (2001). *Effective Integration of IT in Construction*, London.

Cartlidge, D. (2004). *Procurement of Built Assets*. Butterworth Heinemann.

Chappel, l, C. and Feindt, S. (1999). *Analysis of E-Commerce Practice in SMEs*. Kite.

Davis *et al.* (2007). *Knowledge Management for The Quantity Surveying Profession*. Strategic Integration of Surveying Services, FIG.

Department of Trade and Industry (1999). *Building Confidence in Electronic Commerce – A Consultation Document*. HMSO.

Drucker, F. *et al.* (1998). *Harvard Business Review on Knowledge Management*. Harvard Business School Press.

Eadie, R. *et al.* (2010). *A Cross Discipline Comparison of Rankings for e-Procurement Drivers and Barriers Within UK Construction Organisations.* ITcon.

e-Building for Clients (2005). Department of Environment Trade and the Regions.

Frappaolo, C. (2002). *Knowledge Management.* Capstone Publishing.

Garvin, D.A. (1998). *Building a Learning Organisation. Harvard Business Review on Knowledge Management*, Harvard Business School Press.

OGC (2005). *e-Procurement in Action – A Guide to e-Procurement in the Public Sector.* Office of Government Commerce.

Royal Institution of Chartered Surveyors (2005). *E-tendering.* RICS Books.

Royal Institution of Chartered Surveyors (2010). *Contracts in Use.* RICS Books.

SIBIS (2003) *Matching Up to Information Society.* Statistical Indicators Benchmarking the Information Society.

Timmers, P. (1999). *Electronic Commerce – Strategies and Models for Business to Business Trading.* Wiley & Sons Ltd.

Journals

Building (2000). Industry e-commerce to hit £67bn by 2004. 6 October.

Websites

www.ogc.gov.uk/
www.europa.eu.int
www.dti.gov.uk
www.buildingcentretrust.org

Chapter 5

Ethical practice at home and abroad

Ethics

Ethics is an important topic and particularly so for surveyors, who operate in a sector that is generally perceived to have low ethical standards. As if to reinforce the image of the construction industry and contracting as being somewhat dodgy, in April 2008 the Office of Fair Trading (OFT) formally accused 112 construction firms in England of participating in bid rigging on public sector contracts valued at some £200 million. Some of the largest contractors in the UK, including Balfour Beatty, Carillion and Interserve, were alleged to have participated in cartel-type activity in bidding for public sector construction contracts, including schools, universities and hospitals. The focus of the OFT's investigation was on cover pricing, a practice where companies place a high bid for work they have no intention of winning so that they are not left off a client's tender list. In addition, the OFT alleged that a minority of construction companies entered into agreements where the winner of a contract would make a payment of between £2,500 and £60,000 to the unsuccessful bidder, known as a 'compensation payment'. In a damming statement the OFT concluded that it believed the construction industry is rife with malpractice, referring to a previous investigation in 2004 into the roofing industry, when aggregate penalties of £4 million were imposed. In the case of the 2008 allegation the OFT went on, in 2009, to impose fines totalling £129 million on contractors who had been proved to be colluding on bids, including up to £11 million on individual contractors. Subsequently, the companies accused appealed against the fines on the basis that the amounts were disproportionate to the alleged offence.

In 2010 *Building* magazine published the results of a survey conducted by Europe Economics for the OFT where 13 per cent of respondents thought that cover pricing was 'common' or 'appears in most bids' which was the same proportion as in 2008, although the National Federation of Builders dismissed the findings. The same survey found a more tolerant attitude by clients towards contractors who refused to submit bids by not placing them on a blacklist, an approach recommended by OFT guidelines issued in September 2009.

Interestingly, in a survey carried out by the Chartered Institute of Building in 2006, nearly 40 per cent of those questioned viewed the practice of cover pricing as either 'not very corrupt' or 'not corrupt at all', regarding it as the way in which the industry operates. In addition, 41 per cent of respondents admitted offering bribes on one or more occasions. One of the major issues from the CIOB survey is a clear lack of definition of corruption and corrupt practices. The industry is one that depends on personal relationships and yet there is a particular nebulous area in non-cash gifts that range from pens to free holidays.

There is a wide spectrum of research and models on medical ethics-related matters, but comparatively little on business ethics, and even less on ethics and construction and surveying practice. In other sectors in the recent past, corporate/professional malpractice as manifested by the collapse of large corporations such as ENRON and various banking collapses in 2008, followed shortly afterwards by the House of Commons MPs' expenses revelations, led to a public demand for transparency in business, public and professional life. In addition, public figures such as Tiger Woods and John Terry also felt the full force of the media crusade for high ethical and moral standards when millions of pounds of corporate sponsorship disappeared overnight as previously supportive organisations sought to distance themselves from morally tarnished celebrities. Never before has there been such a need for individuals and organisations to be seen to be conducting themselves according to ethical principles.

Ethics and the law

George McKillop in his paper, 'Fraud in construction – follow the money (2009), states: 'I wonder, however, how many people fully understand the true diversity of fraud in construction – not only how endemic it is, but it can affect just about any business.' In a highly critical article he goes on to outline trends in construction fraud giving the following examples:

- The theft and diversion of materials by internal staff.
- Quantity surveyors signing off overpayments in return for kickbacks.
- Construction companies setting up a shell company to invoice a cooperative subcontractor for non-existent services, which in turn is ultimately billed onto his employer's company for signing off.

The legislative framework defining fraud has been confused. The principal statutes currently dealing with corruption are the Public Bodies Corrupt Practices Act 1889, the Prevention of Corruption Act 1906 and the Prevention of Corruption Act 1916. This legislation makes bribery a criminal offence whatever the nationality of those involved if the offer, acceptance or agreement to accept a bribe takes place within the UK's jurisdiction. The

Anti-terrorism, Crime and Security Act 2001 has extended UK jurisdiction to corruption offences committed abroad by UK nationals and incorporated bodies. Commercial bribery is currently covered by the Prevention of Corruption Act 1906 insofar as it relates to bribes accepted by agents. However, the proposed Bribery and Corruption Bill referred to in the Queen's speech in 2009 aims to, according to the then Justice Secretary Jack Straw, 'transform the criminal law on bribery, modernising and simplifying existing legislation to allow prosecutors and the courts to deal with bribery more effectively'. In addition, it was hoped that it will also promote and support ethical practice by encouraging businesses to put in place anti-bribery safeguards to ensure that all employees are aware of the risks surrounding bribery and that adequate systems exist to manage these safeguards. Bribery may include the corruption of a public official as well as commercial bribery, which refers to the corruption of a private individual to gain a commercial or business advantage.

The essential elements of official bribery are:

- giving or receiving;
- a thing of value;
- to influence;
- an official act.

The thing of value is not limited to cash or money. Lavish gifts and entertainment, payment of travel and lodging expenses, payment of credit card bills, 'loans', promises of future employment, interests in businesses are all regarded as bribes if they were given or received with the intent to influence or be influenced. The proposed Bill will make it a criminal offence to give, promise or offer a bribe and to request, agree to receive or accept a bribe either at home or abroad. The measures cover bribery of a foreign public official.

The four principal categories of offences in the proposed Bill will be:

1. Offence of bribing another person.
2. Offences relating to being bribed.
3. Bribery of a foreign public official.
4. The new corporate offence: failure to prevent bribery, whereby a commercial organisation (a corporate or a partnership) could be guilty when:

    - a bribe has been made by a person performing services for or on behalf of the commercial organisation;
    - with the intent to obtain or retain business or other business advantages for the commercial organisation.

It is a defence for the organisation to show that there were adequate procedures in place designed to prevent employees or agents from committing bribery. The penalties on conviction would be the same as for fraud including, in the most serious cases, a sentence of up to ten years' imprisonment following conviction on indictment.

Why is ethics important for surveyors?

Professions can only survive if the public retains confidence in them. Conducting professional activities in an ethical manner is at the heart of professionalism and the trust that the general public has in professions such as the chartered quantity surveyor. One of the principal missions for construction-related institutions like the RICS is to ensure that their members operate to high ethical standards; indeed, ethical standards was a top priority on the RICS *Agenda for Change* (1998). Yet there still appear to be a number of professionals who just don't get it when it comes to ethics, as witnessed by the regular stream of cases that appear before the RICS Disciplinary Panel. In fact the reported cases are just the tip of the iceberg, as many less serious cases brought to the attention of the RICS Professional Conduct Panel are dealt with prior to this stage. For quantity surveyors transparency and ethical behaviour is particularly important as they deal on a day-to-day basis with procurement, contractual arrangements, payments and valuations.

Recently the RICS has published a number of guides/documents to help surveyors find their way through the ethical maze:

- Professional ethics guidance note (2000)
- Professional ethics guidance note (2003) – Case studies
- RICS Core Values (2006)
- RICS Rules of Conduct for Members (2007)
- RICS Rules of Conduct for Firms (2007)
- Fraud in Construction – Follow the Money (2009)
- Fraud in Construction – RICS Guidance (2010).

In addition to the above, the RICS has also published a Help Sheet on Maintaining Professional and Ethical Standards where the behaviour of chartered surveyors in their professional life is characterised as follows:

1. *Act honourably* – never put your own gain above the welfare of your clients or others to whom you have a professional responsibility.
2. *Act with integrity* – be trustworthy in all that you do; never deliberately mislead, whether by withholding or distorting information.
3. *Be open and transparent* in your dealings.
4. *Be accountable* for all your actions and don't blame others if things go wrong.
5. *Know and act within your limitations* and competencies.

6. *Be objective at all times* – never let sentiments or your own interests cloud your judgement.
7. *Always treat others with respect* – never discriminate.
8. *Set a good example* in both public and private behaviour.
9. *Have the courage to make a stand.*
10. *Comply with relevant laws and regulations.*
11. *Avoid conflicts of interest* – declare any potential conflicts of personal or professional interest.
12. *Respect confidentiality* of clients' affairs.

Source: RICS Ethics Help Sheet (2007)

Although the above list appears to be straightforward, things are never quite that simple in practice when matters such as economic survival and competition are added into the mix. The position is even more complicated when operating in countries outside the UK where ideas of ethics may be very different to those expected by the RICS. Ethical behaviour is that which is socially responsible; for example, obeying the law, telling the truth, showing respect for others and protecting the environment.

Concepts of ethics?

Ethical behaviour is developed by people through their physical, emotional and cognitive abilities. People learn ethical behaviour from families, friends, experiences, religious beliefs, educational institutions and the media. Business ethics are shaped by societal ethics.

Ethics are a branch of philosophy that covers a whole range of things that have real importance in one's everyday personal and professional life. Ethics is about:

- right and wrong
- rights and duties
- good and bad
- what goodness itself is
- the way to live a good life
- how people use the language of right and wrong.

In turn ethics tackle some of the fundamentals of life, for example:

- How should people live?
- What should people do in particular situations?

Therefore ethics can provide a moral map, a framework that may be used to find a way through difficult professional issues. Business ethics are about the rightness and wrongness of business practices.

Where do ethics come from?

Where do ethics come from? Have they been handed down in tablets of stone? Some people do think so and philosophers have several answers to this question when they suggest that ethics originate from:

- God – *Supernaturalism*.
- The intuitive moral sense of human beings – *Intuitionism*.
- The example of 'good' human beings – *Consequentialism*.
- A desire for the best for people in each unique situation – *Situation ethics*.

Ethical dilemmas

Instead of trying to arrive at a standard, all-encompassing rule of what is ethical, it is helpful to illustrate the depth and variety of ethics through suitable examples. An ethical dilemma is a situation in which two or more deeply held values come into conflict. In these situations, the correct ethical choice may be unclear.

Perhaps the most common ethical dilemma experienced by quantity surveyors concerns the acceptance of hospitality or gifts from contractors, subcontractors and clients. In the public sector the position is somewhat clearer: employees should not accept gifts as they could be construed as a bribe. The common law offence of bribery involves:

> receiving or offering any undue reward by or to any person whatsoever, in a public office, in order to influence his behaviour in office, and incline him to act contrary to known rules of honesty and integrity.

In addition, Section 1 of the Public Bodies Corrupt Practices Act 1889 makes the bribery of any member, officer or servant of a public body a criminal offence. In particular, the 1889 Act prohibits the corrupt giving or receiving of:

> any gift, loan, fee, reward or advantage whatever, as an inducement to, or reward.

However, in the private sector the guidelines are not so well defined. For example, on completion of a contract for a client, you receive an expensive gift as a mark of appreciation. You are aware that this may be perceived as a way to secure future contracts by the client, but you don't want to appear to be ungrateful. How should you proceed?

A) Accept the gift, pointing out that this does not ensure the awarding of future contracts.

B) Reject the gift, thereby possibly endangering any future business with this client.
C) Enjoy the gift – after all, you deserve it!

(The answer is given at the end of this chapter.)

What is the view of the profession?

During the past few years a number of studies have tried to determine the views of practitioners and students on ethical standards, and there now follows a brief review of some of these studies.

Unethical behaviours, activities and policies

How do you/your organisation measure up? More and more leaders of businesses and other organisations are now waking up to the reality of social

Figure 5.1 Dilemma: What should or ought I do? What is right or wrong? What is good or bad?

responsibility and organisational ethics. Public opinion, unleashed by the internet in particular, is reshaping expectations and standards. Organisational behaviour – good and bad – is more transparent than ever, and injustice anywhere in the world is becoming increasingly visible, and less and less acceptable. As a result reaction to corporate recklessness, exploitation, dishonesty and negligence is becoming more and more organised and potent. So how is it possible to recognise unethical practices within an organisation?

Robert Cooke, one-time director of the Institute of Business Ethics at DePaul University, has identified fourteen danger signs in an organisation that is at risk of unethical behaviour. These are if the organisation:

- normally emphasises short-term revenues over long-term considerations;
- routinely ignores or violates internal or professional codes of ethics;
- always looks for simple solutions to ethical problems and is satisfied with 'quick fixes';
- is unwilling to take an ethical stand when there is a financial cost to the decision;
- creates an internal environment that either discourages ethical behaviour or encourages unethical behaviour;
- usually sends ethical problems to the legal department;
- looks at ethics solely as a public relations tool to enhance its image;
- treats its employees differently from its customers;
- is unfair or arbitrary in its performance-appraisal standards;
- has no procedures or policies for handling ethical problems;
- provides no mechanisms for internal whistle blowing;
- lacks clear lines of communication within the organisation;
- is only sensitive to the needs and demands of the shareholders;
- encourages people to leave their personal ethical values at the office door.

Many commentators have laid the blame for the attitude of corporate at the feet of business schools that have for years put the financial well-being of shareholders at the top of their list of priorities.

Ethics – the business case

Does financial well-being affect attitudes to ethics – is there a business case for ethics? Do ethics add value? There is evidence (see e.g. Moore and Robson, 2002) from other sectors that as organisations' turnover increases their social performance worsens. However, businesses and organisations of all sorts – especially the big high-profile ones – are now recognising that there are solid effects and outcomes driving organisational change. There are now real incentives for doing the right thing, and real disincentives for doing the wrong thing.

As never before, there are huge organisational advantages to behaving ethically, with humanity, compassion and with proper consideration for the world beyond the boardroom and the shareholders:

- *Competitive advantage* – customers are increasingly favouring providers and suppliers who demonstrate responsibility and ethical practices. Failure to do so means lost market share and shrinking popularity, which reduces revenues, profits or whatever other results the organisation seeks to achieve.
- *Better staff attraction and retention* – the best staff want to work for truly responsible and ethical employers. Failing to be a good employer means that good staff members leave, and reduces the likelihood of attracting good new-starters. This pushes up costs and undermines performance and efficiency. Aside from this, good organisations simply cannot function without good people.
- *Investment* – fewer and fewer investors want to invest in organisations that lack integrity and responsibility, both because they don't want the association, and they know that, for all the other reasons listed here, performance will eventually decline, and who wants to invest in a lost cause?
- *Morale and culture* – staff who work in a high-integrity, socially responsible, globally considerate organisation are far less prone to stress, attrition and dissatisfaction. Therefore they are happier and more productive. Happy, productive people are a common feature in highly successful organisations. Stressed, unhappy staff members are less productive, take more time off, need more managing and also take no interest in sorting out the organisation's failings when the whole thing implodes.
- *Reputation* – it takes years, decades, to build organisational reputation, but only one scandal to destroy it. Ethically responsible organisations are far less prone to scandals and disasters. If one does occur, an ethically responsible organisation will automatically know how to deal with it quickly, openly and honestly. People tend to forgive organisations which are genuinely trying to do the right thing. People do not forgive, and are actually deeply insulted by, organisations which fail and then fail again by not addressing the problem and the root cause. Arrogant leaders share this weird delusion that no-one can see what they are up to. Years ago they could hide, but now there is absolutely no hiding place.
- *Legal and regulatory reasons* – soon there will be no choice anyway. All organisations will have to comply with proper ethical and socially responsible standards. These standards and compliance mechanisms

will be global. Welcome to the age of transparency and accountability. It makes sense to change before you are forced to do so.

- *Legacy* – even the most deluded leaders will admit in the cold light of day that they would prefer to be remembered for doing something good, rather than making a pile of money or building a great empire. It is human nature to be good. Humankind would not have survived were this not so. The greedy and the deluded have traditionally been able to persist with unethically irresponsible behaviour because there has been nothing much to stop them, or to remind them that perhaps there is another way. No longer. Part of the reshaping of attitudes and expectations is that making a pile of money, and building a great empire, are becoming stigmatised. What is so great about leaving behind a pile of money or a great empire if it has been at the cost of others' well-being, or the health of the planet? The ethics and responsibility Zeitgeist is fundamentally changing the view of what a lifetime legacy should be and can be. This will change the deeper aspirations of leaders, present and future, who can now see more clearly what a real legacy is.

Ethical research

Each year students from the College of Estate Management in Reading spend two weeks discussing ethical issues in the construction industry in an online asynchronous discussion board. In 2009 the College published 'Ethics for surveyors; An educational dimension'. This study was based on the analysis of an online debate with students and interviews with major property organisations. The research found that:

- High ethical standards are essential to maintaining the reputation of individual surveyors, firms and the wider profession.
- There is evidence of significant ethical issues in surveying practices and a variation in standards between firms and areas of business.
- Corporate culture is more important than professional culture in determining ethical behaviour.
- Firms that operate in countries outside the UK expect the same ethical practices to be applied as in the UK.
- A perception exists among new graduate entrants that the RICS needs to do more to enforce ethical standards.
- Education is ideally placed to raise awareness of ethical issues and to structure and manage the ethics debate.

Ethical decision-making models

The AAA (American Accounting Association) has incorporated the work of Van Hoose and Paradise (1979), Kitchener (1984), Stadler (1986), Haas and Malouf (1989), Forester-Miller and Rubenstein (1992), and Sileo and Kopala (1993) into a practical, sequential, seven-step, ethical decision-making model. A description and discussion of the steps follows.

1. Identify the problem.
   Gather as much information as you can that will illuminate the situation. In doing so, it is important to be as specific and objective as possible. Writing down ideas on paper may help you gain clarity. Outline the facts, separating out innuendoes, assumptions, hypotheses or suspicions. There are several questions you can ask yourself: is it an ethical, legal, professional or clinical problem? Is it a combination of more than one of these? If a legal question exists, seek legal advice.
   Other questions it may be useful to ask yourself are: is the issue related to me and what I am or am not doing? Is it related to a client and/or the client's significant others and what they are or are not doing? Is it related to the institution or agency and their policies and procedures? If the problem can be resolved by implementing the policy of an institution or agency, you can look to the agency's guidelines. It is good to remember that dilemmas you face are often complex, so a useful guideline is to examine the problem from several perspectives and avoid searching for a simplistic solution.
2. Apply the American Counselling Association (ACA) Code of Ethics.
3. After you have clarified the problem, refer to the Code of Ethics (ACA, 2005) to see if the issue is addressed. If there is an applicable standard or several standards and they are specific and clear, following the course of action indicated should lead to a resolution of the problem. To be able to apply the ethical standards, it is essential that you have read them carefully and that you understand their implications.
   If the problem is more complex and a resolution does not seem apparent, then you probably have a true ethical dilemma and need to proceed with further steps in the ethical decision-making process.
4. Determine the nature and dimensions of the dilemma.
   There are several avenues to follow in order to ensure that you have examined the problem in all its various dimensions:

   - Consider the moral principles of autonomy, nonmaleficence, beneficence, justice and fidelity. Decide which principles apply to the specific situation, and determine which principle takes priority for you in this case. In theory, each principle is of equal value, which means that it is your challenge to determine the priorities when two or more of them are in conflict.
   - Review the relevant professional literature to ensure that you are using the most current professional thinking in reaching a decision.
   - Consult with experienced professional colleagues and/or supervisors. As they review with you the information you have gathered, they may

see other issues that are relevant or which provide a perspective you have not considered. They may also be able to identify aspects of the dilemma that you are not viewing objectively.
- Consult your state or national professional associations to see if they can provide help with the dilemma.

5. Generate potential courses of action.
Brainstorm as many possible courses of action as possible. Be creative and consider all options. If possible, enlist the assistance of at least one colleague to help you generate options.
6. Consider the potential consequences of all options and determine a course of action.
Considering the information you have gathered and the priorities you have set, evaluate each option and assess the potential consequences for all the parties involved. Ponder the implications of each course of action for the client, for others who will be affected and for yourself as a counsellor. Eliminate the options that clearly do not give the desired results or cause even more problematic consequences. Review the remaining options to determine which option or combination of options best fits the situation and addresses the priorities you have identified.
7. Evaluate the selected course of action.
Review the selected course of action to see if it presents any new ethical considerations. Stadler (1986) suggests applying three simple tests to the selected course of action to ensure that it is appropriate. In applying the test of justice, assess your own sense of fairness by determining whether you would treat others the same in this situation. For the test of publicity, ask yourself whether you would want your behaviour reported in the press. The test of universality asks you to assess whether you could recommend the same course of action to another counsellor in the same situation.
If the course of action you have selected seems to present new ethical issues, you will need to go back to the beginning and re-evaluate each step of the process. Perhaps you have chosen the wrong option or you may have identified the problem incorrectly.
If you can answer in the affirmative each of the questions suggested by Stadler (thus passing the tests of justice, publicity and universality) and you are satisfied that you have selected an appropriate course of action, then you are ready to move on to implementation.
8. Implement the course of action.
Taking the appropriate action in an ethical dilemma is often difficult. The final step involves strengthening your ego to allow you to carry out your plan. After implementing your course of action, it is good practice to follow up on the situation to assess whether your actions had the anticipated effect and consequences.

Source: American Accounting Organisation.

Another commonly used ethical decision-making model is the American Accounting Association.

The American Accounting Association

Again, similar to the previous AAA model, this process poses a number of questions.

Step 1: What are the facts of the case?
Step 2: What are the ethical issues in the case?
Step 3: What are the norms, principles and values related to the case?
Step 4: What are the alternative courses of action?
Step 5: What is the best course of action that is consistent with the norms, principles and values identified in Step 3?
Step 6: What are the consequences of each possible course of action?
Step 7: What is the decision?

The Laura Nash model

Laura Nash was a Senior Research Fellow at Harvard Business School in 1981 when she drafted a series of twelve questions to help when making difficult ethical decisions as follows:

1. Have you defined the problem accurately?
2. How would you define the problem if you stood on the other side of the fence?
3. How did this situation occur in the first place?
4. To whom and to what do you give your loyalty as a person and as a member of an organisation?
5. What is your intention in making this decision?
6. How does the intention compare with the probable results?
7. Whom could your decision injure?
8. Can you discuss the problem with the affected parties before you make a decision?
9. Are you confident that your position will be as valid over a long period of time as it seems now?
10. Could you disclose without qualms your decision or action to your boss, the head of your organisation, your colleagues, your family, the person you most admire or to society as a whole?
11. What is the symbolic potential of your action if understood? If misunderstood?
12. Are there circumstances when you would allow exceptions to your stand? What are they?

Of all the decision-making models that have been developed. the Nash model would appear to be one of the more respected. Other models include:

- Tucker's 5 Question Model
- The Mary Guy Model(1990)
- The Rion Model (1990)
- The Langenderfer and Rockness Model (1990).

Sustainability and ethics

The question of sustainability is discussed in Chapters 3 and 8.

An international perspective

If ethical dilemmas are problematic on home turf, the problems become magnified and more diverse for construction professionals working outside of the UK.

Transparency International is an organisation that seeks to provide reliable quantitative diagnostic tools regarding levels of transparency and corruption, both at global and local levels. Each year the company publishes a Corruption Perceptions Index that ranks countries and illustrates how countries compare with each other in terms of perceived levels of public sector corruption. Table 5.1 illustrates the results of the 2009 transparency survey. There are 180 counties in the full survey that can be accessed at www.transparency.org.

By its very nature there is little concrete evidence on the levels of bribery and corruption in the world market but companies based in emerging economic giants, such as China, India and Russia, are perceived to routinely engage in bribery when doing business abroad, according to Transparency International's 2008 Bribe Payers Index (BPI) (see Table 5.2).

Belgium and Canada shared first place in the 2008 BPI with a score of 8.8 out of a very clean 10, indicating that Belgian and Canadian firms are seen as least likely to bribe abroad. The Netherlands and Switzerland shared

Table 5.1 Results of Transparency International Survey (2009)

Country	*Rank*	*CPI score*	*Country*	*Rank*	*CPI score*
New Zealand	1	9.4	Malaysia	56	4.5
Hong Kong	12	8.2	Italy	63	4.3
Germany	14	8.0	India	84	3.4
United Kingdom	17	7.7	Nigeria	130	2.5
Qatar	22	7.0	Russia	146	2.2
Estonia	27	6.6	Venezuela	162	1.9
South Africa	56	4.5	Somalia	180	1.1

third place on the Index, each with a score of 8.7. At the other end of the spectrum, Russia ranked last with a score of 5.9, just below China (6.5), Mexico (6.6) and India (6.8). Interestingly the BPI also shows public works and construction companies to be the most corruption-prone when dealing with the public sector, and most likely to exert undue influence on the policies, decisions and practices of governments.

The BPI provides evidence that a number of companies from major exporting countries still use bribery to win business abroad, despite awareness of its damaging impact upon corporate reputations and ordinary communities.

The Bribe Payers Survey, which serves as the basis for the BPI, also looks at the likelihood of firms in nineteen specific sectors to engage in bribery. In the first of two new sectoral rankings, companies in public works contracts and construction; real estate and property development; oil and gas; heavy manufacturing; and mining were seen to bribe officials most frequently. The cleanest sectors, in terms of bribery of public officials, were identified as information technology, fisheries and banking and finance.

A second sectoral ranking evaluates the likelihood of companies from the nineteen sectors to engage in state capture, whereby parties attempt to wield undue influence on government rules, regulations and decision-making through private payments to public officials. Public works contracts and construction; oil and gas; mining; and real estate and property development were seen as the sectors whose companies were most likely to use legal or illegal payments to influence the state. The banking and finance sector is seen to perform considerably worse in terms of state capture than in willingness to bribe public officials, meaning that its companies may exert considerable undue influence on regulators, a significant finding in light of the ongoing global financial crisis. The sectors where companies are seen as least likely to exert undue pressure on the public policy process are agriculture, fisheries and light manufacturing; however, it was estimated that in 2005 corruption was costing the construction industry worldwide £3.75 billion per annum.

While most of the world's wealthiest countries already subscribe to a ban on foreign bribery under the OECD Anti-Bribery Convention, there is little awareness of the Convention among the senior business executives interviewed in the Bribe Payers Survey. Governments play a key role in ensuring that foreign bribery is stopped at the source – and by making good on commitments to prevent and prosecute such practices.

Table 5.2 shows the 2008 BPI results along with additional statistical information that indicates the level of agreement among respondents about the country's performance, and the precision of the results. Scores range from 0 to 10, indicating the likelihood of firms headquartered in these countries to bribe when operating abroad. The higher the score for the country, the lower the likelihood of companies from this country to engage in bribery when conducting business abroad.

Table 5.2 Transparency International's Bribe Payers Index (2008)

Rank	*Country*	*BPI 2008 score*
1	Belgium	8.8
1	Canada	8.8
3	Netherlands	8.7
3	Switzerland	8.7
5	Germany	8.6
5	United Kingdom	8.6
5	Japan	8.6
8	Australia	8.5
9	France	8.1
9	Singapore	8.1
9	United States	8.1
12	Spain	7.9
13	Hong Kong	7.6
14	South Africa	7.5
14	South Korea	7.5
14	Taiwan	7.5
17	Italy	7.4
17	Brazil	7.4
19	India	6.8
20	Mexico	6.6
21	China	6.5
22	Russia	5.9

In 2009 the FIEC issued a statement in its annual report that the World Bank and the European Union had singled out the construction industry as a sector which, according to their perception, is particularly prone to unethical business practices. As a result a joint FIEC/EIC Working Group of Ethics was established which in 2009 published a *Statement on Corruption Prevention in the Construction Industry* that basically denounced corruption in the construction sector and briefly outlined a code of conduct to promote transparency.

An ethics charter

One step in raising the awareness of ethics within an organisation is by drawing up an ethics charter. Some guidelines on implementing this are set out below:

1. Get endorsement from the Board. Corporate values and ethics are matters of governance.
2. Find a champion. It is good practice to set up a board-level (ethics or corporate responsibility) committee, preferably chaired by a non-executive director, or to assign responsibility to an existing committee

(such as Audit or Risk). A senior manager will need to be responsible for the development of the policy and code, and the implementation of the ethics programme.

3. Understand the purpose. It is important to clarify the relationship between and to understand your organisation's approach to corporate responsibility, ethics, compliance and corporate social responsibility strategies.
4. Find out what bothers people. Merely endorsing an external standard or copying a code from another organisation will not suffice. It is important to find out on what topics employees require guidance, to be clear what issues are of concern to stakeholders and what issues are material to your business activities, locations and sector.
5. Be familiar with external standards and good practice. Find out how other companies in your sector approach ethics and corporate responsibility. Understand what makes an effective policy, code and programme from the point of view of the business, the staff and other stakeholders. How will you embed your code into business practice?
6. Monitoring and assurance. Consider how the success of the policy will be monitored and to whom the business will be accountable regarding its ethical commitments. How will you know it is working? What are the key indicators/measures of an ethical culture for your organisation?
7. Try it out first. The draft code needs piloting – perhaps with a sample of employees drawn from all levels and different locations.
8. Review. Plan a process of review that will take account of changing business environments, strategy, stakeholder concerns and social expectations, new standards and strengths and weakness in your ethical performance.

Another major ethical dilemma that sometimes has to be faced is whether to expose malpractice within an organisation for whom you work to the public domain, a practice that has become known as whistle blowing.

Whistle blowing

The definition of whistle blowing may be said to be speaking out to the media or the public on malpractice, misconduct, corruption or mismanagement witnessed in an organisation. Whistle blowing occurs when a worker raises a concern about danger or illegality that affects others, for example, members of the public. Whistle blowing is usually undertaken on the grounds of morality or conscience, or because of a failure of business ethics on the part of the organisation being reported. Put at its simplest, whistle blowing occurs when an employee or worker provides certain types of information, usually to the employer or a regulator, which have come to their attention through work. The whistle blower is usually not directly, personally affected by the danger or illegality.

Whistle blowing is not for the faint hearted, as whistle blowers have often been the subject of victimisation, threats, bullying and dismissal. As mentioned earlier there are many case studies on ethical/whistle-blowing issues in other sectors, particularly the health sector, but far fewer in construction and related professions. There is however one case that came to prominence in 2005 when production manager Alan Wainwright, an ex-employee of Haden Young, alleged that the company was operating a policy of keeping a blacklist database. The database, it was alleged, contained over 500 names of people and companies considered to be disruptive or militant and who should not be hired. In 2005 Wainwright went public on the blacklist issue and shortly thereafter left the company and subsequently failed after 150 job applications to secure a new job, leading to the conclusion that he himself had been placed on a blacklist. Finally, after 200 applications, Wainwright now works as a concert ticket buyer. The consequences for Wainwright were losing his job, having no income, stress and fear of eviction from his home. In May 2009, Lord Mandelson vowed to introduce legislation to outlaw the compilation and operation of blacklists.

While waiting for the promised legislation the key piece of existing whistle-blowing legislation is the Public Interest Disclosure Act 1998 (PIDA) which applies to almost all workers and employees who ordinarily work in Great Britain. The provisions introduced by the Public Interest Disclosure Act 1998 protect most workers from being subjected to a detriment by their employer. Detriment may take a number of forms, such as denial of promotion, facilities or training opportunities which the employer would otherwise have offered. Under the provisions of the PIDA certain kinds of disclosure qualify for protection ('qualifying disclosures'). Qualifying disclosures are disclosures of information which the worker reasonably believes tend to show that one or more of the following is either happening now, took place in the past or is likely to occur in the future:

- a criminal offence;
- the breach of a legal obligation;
- a miscarriage of justice;
- a danger to the health or safety of any individual;
- damage to the environment;
- deliberate covering up of information tending to show any of the previous five offences.

It should be noted that in making a disclosure the worker must have reasonable belief that the information disclosed tends to show one or more of the offences or breaches listed above ('a relevant failure'). The belief need not be correct – it might be discovered subsequently that the worker was in fact wrong – but the worker must show that he held the belief, and that it was a reasonable belief in the circumstances at the time of disclosure.

Employment tribunal statistics show that the total number of people using whistle-blowing legislation, which aims to protect workers from victimisation if they have exposed wrongdoing, increased from 157 cases in 1999 to 1,791 ten years later. These figures, compiled for the first time, will increase fears among campaigners that whistle blowers are being deliberately undermined or removed from their workplace, despite repeated promises to protect them.

Whistle-blowing procedures

Attitudes towards whistle blowing have evolved considerably during the past fifty years in the early days of the 'organisation man' ethos where loyalty to the company was the ruling norm, to the present time when public outrage about corporate misconduct has created a more auspicious climate for whistle blowing.

Companies had broad autonomy in employee policies and could fire an employee at will, even for no reason. Employees were expected to be loyal to their organisations at all costs. Among the few exceptions to this rule were unionised employees, who could only be fired for 'just cause', and government employees because the courts upheld their constitutional right to criticise agency policies. In private industry, few real mechanisms for airing grievances existed, although, for example, IBM claimed from its earliest days to have an effective open-door policy that allowed employees to raise any issue. In part due to this lack of protection for whistle blowers, problems were often concealed rather than solved. Probably the most clear example was in asbestos manufacturing, where the link to lung disease was clearly established as early as 1924 but actively suppressed by company officials for fifty years. In the late 1960s it was commonplace in the UK to see joiners cutting and drilling asbestos sheets on site, without any form of protection, before fixing it below board flooring as a fire barrier.

Any whistle-blowing policy should aim to:

- encourage staff to feel confident in raising serious concerns and to question and act upon concerns about practice;
- provide avenues for staff to raise those concerns and receive feedback on any action taken;
- ensure that staff receive a response to concerns and that you are aware of how to pursue them if not satisfied;
- reassure staff members that they will be protected from possible reprisals or victimisation if you have a reasonable belief that any disclosure has been made in good faith.

Non-UK markets

One of the many factors having to be faced when considering expansion into non-UK markets is the question of differing ethical standards and how these should be navigated. During the past five years the RICS has increased its global presence, including the appearance of several European and US universities and higher education establishments in the RICS Partnership Programme. However, all eyes are on China, where the economy is growing at 9 per cent per annum compared with the average for the Euro Zone of 1.8 per cent and a 26 per cent year-on-year increase in infrastructure investment such as bridges, factories and power plants.

The multicultural team

The *Le Monde* cartoon featured in Figure 5.2 illustrates most quantity surveyors' perception or even experience of working in or with multicultural teams. Quantity surveyors have proved themselves to be adept in a diverse range of skills, often over and above their technical knowledge, with which they serve the needs of their clients. However, when operating in an international environment these skills and requirements are complicated by the added dimension of a whole series of other factors, including perhaps the most influential – cultural diversity. Companies operating at an international

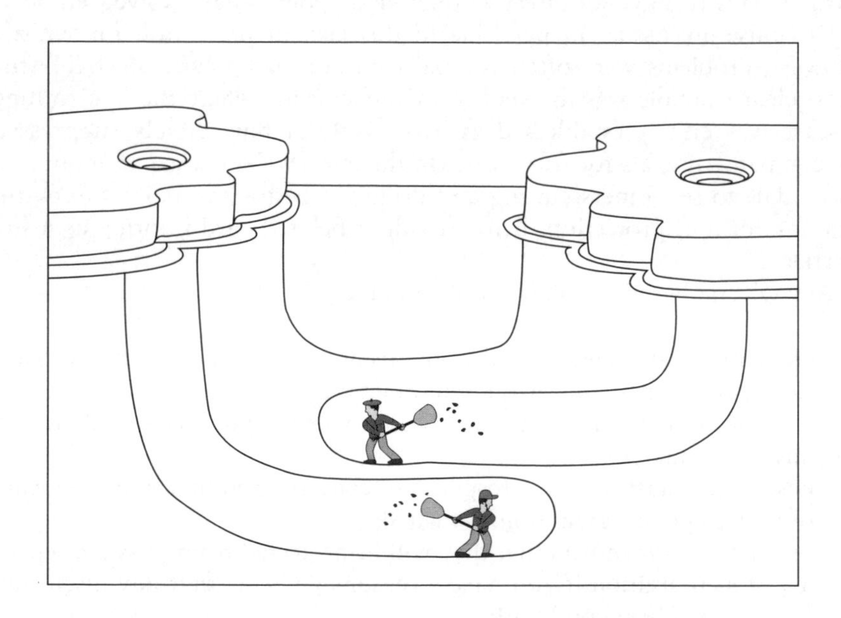

Figure 5.2 Making ends meet

Source: Serguei – Le Monde.

level in many sectors have come to realise the importance of a good understanding of cultural issues and the impact they have upon their business operations. In an increasingly global business environment in which the RICS is constantly promoting the surveyor as a global player (for example, the RICS Global Manifesto), it is a fact that the realisation of the importance and influence of cultural diversity is still lacking in many organisations seeking to expand their business outside of the UK.

Today for many quantity surveyors, international work is no longer separated from the mainstream surveying activity; EU Procurement Directives, GATT/GPA (Government Procurement Agreement of the World Trade Organisation) and so on are bringing an international dimension to the work of the property professionals. Consultants from the UK are increasingly looking to newer overseas markets such as Europe and regions where they have few traditional historical connections, such as South East Asia and China. Consultants must compete with local firms in all aspects of their services, including business etiquette, market knowledge and fees but above all, delivering added value in order to succeed. With the creation of the single European market in 1993, many UK firms were surprised that European clients were not interested in the novelty value of using UK professionals, but continued to award work on the basis of best value for money. As pointed out in Chapter 1, the UK construction industry and its associated professions have ploughed a lonely furrow for the past 150 years or so as far as the status and nature of the professions, procurement and approach to design are concerned, and it could be argued that this baggage makes it even more difficult to align with and/or adapt to overseas markets. Certainly companies like the French giant Bouygues, with its multidisciplinary *bureau d'études techniques,* have a major advantage in the international markets owing to their long-established capability to re-engineer initial designs in-house and present alternative technical offers.

Outside of Europe, the US, for example, has seen a decline in the performance of the US construction industry in international markets, a trend that has been attributed in part to its parochial nature in an increasingly global market. Internally, strong trade unions exercise a vice-like grip on the American construction industry.

The construction industry, in common with many other major business sectors, has been dramatically affected by market globalisation. Previous chapters have described the impact of the digital economy upon working practices; multinational clients such as Coca-Cola and BP demand global solutions to their building needs, and professional practices as well as contractors are forging international alliances (either temporary or permanent) in order to meet demand. It is a fast-moving and highly competitive market, where big is beautiful and response time is all-important. The demands placed on professional consultants with a global presence are high, particularly in handling unfamiliar local culture, planning regimes and procurement practice, but the reward is greater consistency of workload for consultants and contractors alike. In an increasingly competitive environment, the companies that are operating at an international level in many sectors have come to realise that a good awareness and understanding of cultural issues is essential to their international business performance. Closeness and interrelationships within the international business community are hard to penetrate without acceptance as an insider, which can only be achieved with cultural and social understanding. In order to maintain market share, quantity surveyors need to tailor their marketing strategies, for example, to take account of the different national cultures. Although some differences turn out to be ephemeral, when exploring international markets there is often a tightrope that has to be walked between an exaggerated respect, which can appear insulting, and a crass insensitivity, which is even more damaging. It comes as no surprise that cultural diversity has been identified as the single greatest barrier to business success.

It is no coincidence that the global explosion occurred just as the e-commerce revolution arrived, with its 365-days-a-year/twenty-four-hours-a-day culture, allowing round-the-clock working and creating a market requiring international expertise backed by local knowledge and innovative management systems. Although it could be argued that in an e-commerce age cultural differences are likely to decrease in significance, they are in fact still very important, and remain major barriers to the globalisation of e-commerce. These differences extend also to commercial practice. Even in a digital economy an organisation still needs to discover and analyse a client's values and preferences, and there is still a role for intermediaries such as banks, trading companies, international supply chain managers and chambers of commerce in helping to bridge differences in culture, language and commercial practice. In an era of global markets purists could perhaps say that splitting markets into European and global sectors is a contradiction. However, Europe does have its own unique features, not least its public procurement directives, physical link and proximity to the UK, and for some states a single currency. Therefore, this chapter will first consider the European market, before looking at other opportunities.

Europe

During the early 1990s Euromania broke out in the UK construction industry, and 1 January 1993 was to herald the dawn of new opportunity. It was the day the remaining physical, technical and trade barriers were removed across Europe, and from now on Europe and its markets lay at the UK construction industry's feet. Optimism was high within the UK construction industry – after all, it seemed as though a barrier-free Europe with a multibillion pound construction-related output (€1305 billion in 2009 according to the European Construction Industry Federation; see Table 5.3) was the

Table 5.3 EU top construction-related output (2009)

Country	*Billion*
Germany	251
Spain	193
France	172
United Kingdom	155
Italy	153
Netherlands	61
Belgium	35
Austria	33
Denmark	32
Ireland	30
Finland	30
Sweden	27
Czech Republic	20
Portugal	20
Greece	15
Romania	14
Hungary	9

Source: FIEC.

solution to the falling turnover in the UK. Almost every month conferences were held on the theme of how to exploit construction industry opportunities in Europe. So, more than a decade later, has the promise been turned into a reality? There follows a review of developments in Europe, with an examination of the procurement opportunities for both the public and private sectors.

Of the total European construction output, approximately:

- 21 per cent was in civil engineering;
- 20 per cent was in new house building;
- 31 per cent was in non-residential (offices, hospitals, hotels, schools, industrial buildings);
- 28 per cent was in rehabilitation and maintenance.

European public procurement law

Procurement in the European public sector involves governments, utilities (i.e. entities operating in the water, energy and transport sectors) and local authorities purchasing goods, services and works over a wide range of market sectors, of which construction is a major part. For the purposes of legislation, public bodies are divided into three classes:

1. Central government and related bodies (e.g. NHS Trusts).
2. Other public bodies (e.g. local authorities, universities).
3. Public utilities (e.g. water, electricity, gas, rail).

Public procurement is different from private business transactions in several aspects; the procedures and practices are heavily regulated and, while private organisations can spend their own budgets more or less as they wish (with the agreement of their shareholders), public authorities receive their budgets from taxpayers and therefore have a responsibility to obtain value for money, traditionally based on lowest economic cost. However, in recent years the clear blue water between private and public sectors has disappeared rapidly with the widespread adoption of public/private partnerships and the privatisation of what were once publicly owned utilities or entities.

The Directives – theory and practice

The EU Directives provide the legal framework for the matching of supply and demand in public procurement. A Directive is an instruction addressed to the EU Member States to achieve a given legislative result by a certain deadline. This is usually done by transposing the terms of the Directive into national legislation. The European public procurement regulatory framework was established by the public procurement Directives 93/36/EEC, 93/37/EEC and 92/50/EEC for supplies, works and services and Directive 93/38/EEC for utilities, which, together with the general principles enshrined in the Treaty of Rome (1957), established the following principles for cross-border trading (references apply to the Treaty of Rome):

- A ban on any discrimination on the grounds of nationality (Article 6).
- A ban on quantitative restrictions on imports and all measures having equivalent effect (Articles 30 to 36).
- The freedom of nationals of one Member State to establish themselves in another Member State (Articles 52 *et seq.*) and to provide services in another Member State (Articles 59 *et seq.*).

Enforcement Directives (89/665EEC and 92/13EEC) were added in 1991 in order to deal with breaches and infringements of the system by Member States.

The quantity surveyor and EU public procurement

How is the quantity surveyor likely to come into contact with the European public procurement system? The following scenarios are discussed:

- A surveyor working within a public body (contracting authority) and dealing with a works contract.
- A surveyor in private practice wishing to bid for work in Europe as a result of a service contract announcement.

A surveyor within a public body

A quantity surveyor working within a body governed by public law (if in doubt, a list of European bodies and categories of bodies is given in the Directives) should be familiar with procedures for compliance with European public procurement law. The Directives lay down thresholds above which it is mandatory to announce the contract particulars. The *Official Journal of the European Communities* is the required medium for contract announcements and is published five times a week, containing up to 1,000 notices covering every imaginable contract required by central and local government and the utilities – from binoculars in Barcelona to project management in Porto. Major private sector companies also increasingly use the *Official Journal* for market research. The current thresholds (effective from January 2010) for announcements in the Official Journal are given in Table 5.4. and 5.5.

DGXV actively encourages contracting authorities and entities to announce contracts that are below threshold limits. Information on these impending

Table 5.4 Public contracts

	Supplies	*Services*	*Works*
Entities listed in Schedule I[a]	£101,323 (€125,000)	£101,323 (€125,000)	£3,927,260 (€4,845,000)
Other public sector contracting authorities	£156,442 (€193,000)	£156,442 (€193,000)	£3,927,260 (€4,845,000)
Prior Information Notices (Regulation 11)	£607,935 (€750,000)	£607,935 (€750,000)	£3,927,260 (€4,845,000)
Small lots (Regulation 8 (12))	£64,846 (€80,000)	£64,846 (€80,000)	£810,580 (€1,000,000)

Note
a Schedule 1 of the Public Contracts Regulations 2006 lists central government bodies subject to the WTO GPA. These thresholds will also apply to any successor bodies.

Table 5.5 Utilities contracts

	Supplies	*Services*	*Works*
Threshold (regulation 11)	£313,694 (€387,000)	£313,694 (€387,000)	£3,927,260 (€4,845,000)
Periodic Indicative Notices (Regulation 15)	£607,935 (€750,000)	£607,935 (€750,000)	£3,927,260 (€4,845,000)
Small lots (Regulation 11(9))	£64,846 (€80,000)	£64,846 (€80,000)	£810,580 (€1,000,000)

Source: Office of Government Commerce (OGC).

Note: All figures exclude VAT.

tenders is published by the European Commission in the *Official Journal of the European Communities*, often otherwise known as the *OJEU* , which is available free of charge electronically at europa.eu.

The Directive also clarifies existing law in areas such as the selection of tenderers and the award of contracts, bringing the law as stated into line with judgments of the European Court of Justice.

The EU procurement procedure

The *OJEU* announcement procedure involves three stages:

1. Prior information notices (PINs) or indicative notices.
2. Contract notices.
3. Contract award notices (CANs).

Examples of these notices may be found in Annex IV of the Directive.

- *A prior information notice,* or PIN, that is not mandatory, is an indication of the essential characteristics of a works contract and the estimated value. It should be confined to a brief statement, and posted as soon as planning permission has been granted. The aim is to enable contractors to schedule their work better and to allow contractors from other member states the time to compete on an equal footing.
- *Contract notices* are mandatory and must include the award criteria, which may be based upon either the lowest price or the most economically advantageous tender, specifying the factors that will be taken into consideration. Once drafted, the notices are published five times a week, via the Publications Office of the European Commission in Luxembourg in the *Official Journal* via the Tenders Electronic Daily (TED) database, and translated into the official languages of the community, all costs being borne by the community. TED is updated twice weekly and may be accessed through the Commission's website at http://simap.eu.int. Extracts from TED are also published weekly in the trade press. In order to give all potential contractors a chance to tender for a contract, the Directives lay down minimum periods of time to be allowed at various stages of the procedure – for example, in the case of Open Procedure this ranges from thirty-six to fifty-two days from the date of dispatch of the notice for publication in the *Official Journal*. Restricted and negotiated procedures have their own time limits. These timescales should be greatly reduced with the wide-scale adoption of electronic procurement.
- *Contract award notices* inform contractors about the outcome of the procedure. If the lowest price was the standard criterion, this is not difficult to apply. If, however, the award was based on the 'most economically advantageous tender', further clarification is required to explain

the criteria (e.g. price, period for completion, running costs, profitability and technical merit), listed in descending order of importance. Once established, the criteria should be stated in the contract notices or contract documents.

Award procedures

The surveyor must decide at an early stage which award procedure is to be adopted. The following general criteria apply:

- minimum number of bidders must be five for the restricted procedure and three for the negotiated and competitive dialogue procedures;
- contract award is made on the basis of lowest price or most economically advantageous tender (MEAT);
- contract notices or contract documents must provide the relative weighting given to each criterion used to judge the most economically advantageous tender and, where this is not possible, award criteria must be stated in descending order of importance;
- MEAT award criteria may now include environmental characteristics (e.g. energy savings, disposal costs), provided that these are linked to the subject matter of the contract.

The choices are as follows:

- *Open procedure*, which allows all interested parties to submit tenders.
- *Restricted procedure*, which initially operates as the open procedure but then the contracting authority only invites certain contractors, based on their standing and technical competence, to submit a tender. Under certain circumstances, for example, extreme urgency, this procedure may be accelerated.
- *Negotiated procedure*, in which the contracting authority negotiates directly with the contractor of its choice. This is used in cases where it is strictly necessary to cope with unforeseeable circumstances, such as an earthquake or flood. It is most commonly used in PPP models in the UK.
- *Competitive dialogue*, the introduction of this procedure addresses the need to grant, in the opinion of the European Commission, contracting authorities more flexibility to negotiate on public/private partnership (PPP) projects. Some contracting authorities have complained that the existing procurement rules are too inflexible to allow a fully effective tendering process. Undoubtedly, the degree of concern has depended largely on how a contracting authority has interpreted the procurement rules, as there are numerous examples of public/private partnership projects which have been successfully tendered since the introduction of

the public procurement rules using the negotiated procedure. However, the European Commission recognised the concerns being expressed, not only in the UK but across Europe, and it has sought to introduce a new procedure which will accommodate these concerns. In essence, the new competitive dialogue procedure permits a contracting authority to discuss bidders' proposed solutions with them before preparing revised specifications for the project and going out to bidders asking for modified or upgraded solutions. This process can be undertaken repeatedly until the authority is satisfied with the specifications that have been developed. Some contracting authorities are pleased that there is to be more flexibility to negotiations; however for bidders this reform does undoubtedly mean that tendering processes could become longer and more complex. This in turn would lead to more expense for bidders and could pose a threat to new entrants to the PPP market as well as existing players. According to the Commission's DGXV Department the introduction of this procedure will enable:

- dialogue with selected suppliers to identify and define solutions to meet the needs of the procuring body;
- awards to be made only on the basis of the most economically advantageous.

In addition:

- All candidates and tenderers must be treated equally and commercial confidentiality must be maintained unless the candidate agrees that information may be passed on to others.
- Dialogue may be conducted in successive stages. Those unable to meet the need or provide value for money, as measured against the published award criteria, may drop out or be dropped, although this must be conveyed to all tenderers at the outset.
- Final tenders are invited from those remaining on the basis of the identified solution or solutions.
- Clarification of bids can occur pre- and post-assessment provided that this does not distort competition.

To summarise, therefore, the competitive dialogue procedure is, according to the Commission, to be used in cases where it is difficult to access what would be the best technical, legal or financial solution owing to the market for such a scheme or the project being particularly complex. However, the competitive dialogue procedure leaves many practical questions over its implementation; for example:

- The exceptional nature of the competitive dialogue and its hierarchy with other award procedures.

- The discretion of the contracting authorities to initiate the procedure. Who is to determine the nature of a particular complex project?
- The response of the private sector, with particular reference to the high bid costs.
- The overall value for money.
- The degree of competition achieved, since there is great potential for post-contract negotiations.

Electronic tendering

Electronic auctions

The internet is making the use of electronic auctions increasingly more attractive as a means of obtaining bids in both public and private sectors; indeed it can be one of the most transparent methods of procurement. At present electronic auctions may be used in both open and restricted framework procedures. The system works as follows:

- The framework (i.e. of the selected bidders) is drawn up.
- The specification is prepared.
- The public entity then establishes the lowest price award criterion (e.g. with a benchmark price as a starting point for bidding).
- Reverse bidding on a price then takes place, with framework organisations agreeing to bid openly against the benchmark price.
- Prices/bids are posted up to a stated deadline.
- All bidders see the final price.

Technical specifications

At the heart of all domestic procurement practice is compliance with the technical requirements of the contract documentation in order to produce a completed project that performs to the standards of the brief. The project must comply with national standards or be compatible with existing systems and technical performance. The task of achieving technical excellence becomes more difficult when there is the possibility of the works being carried out by a contractor who is unfamiliar with domestic conventions and is attempting to translate complex data into another language. It is therefore very important that standards and technical requirements are described in clear terms with regard to the levels of quality, performance, safety, dimensions, testing, marking or labelling, inspection, and methods or techniques of construction. References should be made to the following:

- *A Standard:* a technical specification approved by a recognised standardising body for repeated and continuous application.

- *A European Standard:* a standard approved by the European Committee for Standardisation (CEN).
- *European technical approval:* a favourable technical assessment of the fitness for use of a product, issued by an approval body designated for the purpose (sector-specific information regarding European technical approval for building products is provided in Directive 89/106/EEC).
- *Common technical specification:* a technical specification laid down to ensure uniform application in all member states, which has been published in the *Official Journal.*
- *Essential requirements:* requirements regarding safety, health and certain other aspects in the general interests that the construction works must meet.

Given the increased complexity of construction projects, the dissemination of accurate and comprehensive technical data is gaining in importance. It is therefore not surprising that the Commission is concerned that contracting authorities are, either deliberately or otherwise, including discriminatory requirements in contract documents. These include the following:

- lack of reference to European standards;
- application of technical specifications that give preference to domestic production;
- requirements of tests and certification by a domestic laboratory.

The result of this is in direct contravention of Article 30 of the Treaty of Rome, and effectively restricts competition to domestic contractors. In an attempt to reduce the potential problems outlined above, the EU has embarked on a campaign to encourage contracts to be based upon an output or performance specification, which removes the need for detailed and prescriptive documentation.

The Code of Conduct for European surveyors

The European Commission has encouraged all professions to develop a Code of Conduct and in 2007 published guidance on the development of such codes. The aim of the codes is to encourage trust among European consumers who may otherwise consider that organisations with only national codes could lead to mistrust as well as making penetration of European markets more difficult. The responsibility for the development of a Code of Conduct for European surveyors has been given to the Comite de Liaison des Géomètres Européens-GeometerEuropas. The Code is expected to be ratified in 2010 and may be viewed at www.cige.eu.

Public procurement beyond Europe

There are no multilateral rules governing public procurement. As a result, governments are able to maintain procurement policies and practices that are trade distortive. That many governments wish to do so is understandable; government purchasing is used by many as a means of pursuing important policy objectives that have little to do with economics – social and industrial policy objectives rank high among these. The plurilateral Government Procurement Agreement (GPA) partially fills the void. The GPA is based upon the GATT provisions negotiated during the 1970s, and is reviewed and refined at meetings (or rounds) by ministers at regular intervals. Its main objective is to open up international procurement markets by applying the obligations of non-discrimination and transparency to the tendering procedures of government entities. It has been estimated that market opportunities for public procurement increased tenfold as a result of the GPA. The GPA's approach follows that of the European rules. The Agreement establishes a set of rules governing the procurement activities of member countries and provides for market access opportunities. It contains general provisions prohibiting discrimination as well as detailed award procedures. These are quite similar to those under the European regime, covering both works and other services involving, for example, competition, the use of formal tendering and enforcement, although the procedures are generally more flexible than under the European rules. However, the GPA does have a number of shortcomings. First, and perhaps most significantly, its disciplines apply only to those World Trade Organisation members that have signed it. The net result is a continuing black hole in multilateral WTO rules that denies access or provides no legal guarantees of access to billions of dollars of market opportunities in both the goods and services sector. The current parties are the European Union, Aruba, Norway, Canada, Israel, Japan, Liechtenstein, South Korea, the USA, Switzerland and Singapore.

Developments in public procurement

As in the private sector, information technology is the driving force in bringing efficiency and added value to procurement. However, despite the many independent research projects that have been undertaken by the private sector, the findings cannot simply be lifted and incorporated into the public sector due to the numerous UK and European Community regulations that must be adhered to. Notwithstanding these potential problems, the UK government has set an ambitious target for the adoption of e-tendering in the public sector.

Of all strands of the e-business revolution, it is e-procurement that has been the most broadly adopted, has laid claim to the greatest benefits and accounts for the vast majority of electronic trading. A survey carried out on behalf of the EU in 2000 showed that, of the existing electronic procurement systems in use, building and construction was offered by all of them and was the top-ranked sector, with a usage rate of 72 per cent.

Europe and beyond

The effect of culture on surveyors operating in international markets

As discussed at the opening of this chapter, culture can be a major barrier to international success. Culture must first be defined and then analysed so that it can be managed effectively; thereafter there is the possibility of modelling the variables as an aid to business. A business culture does not change quickly, but the business environment from which it is derived and with which it constantly interacts is sometimes subject to radical and dramatic change. The business culture in a particular country grows partly out of what could be called the current business environment of that country. Yet business culture is a much broader concept, because alongside the impulses that are derived from the current business environment there are historical examples of the business community. For example, as discussed in Chapter 1, the 1990 recession saw widespread hardship, particularly in the UK construction industry. There have been many forecasts of doom during the early twenty-first century from analysts drawing comparisons between the state of current business with that in 1990, when record output, rising prices and full employment were threatening to overheat the economy as well as construction – can there be many quantity surveying practices in the UK that are not looking over their shoulders to see if and when the next recession is coming? Table 5.6 outlines a sample of the responses by 1,500 European companies questioned during a study into the effects of culture on business.

Table 5.6 Sample of responses from 1,500 European companies as to effects of culture on business

China	Cultural differences are as important as an understanding of Asian or indeed other foreign languages.
Far East	One needs to know etiquette/hierarchical structure/manner of conduct in meetings.
Germany	Rigid approach to most operational procedures.
Middle East	Totally different culture – time, motivation, responsibility.
Russia	Inability to believe terms and conditions as stated really are what they are stated to be.
SE Asia	Strict etiquette of business in South Korea and China can be a major problem if not understood.
France	Misunderstandings occurred through misinterpretation of cultural differences.

So what is culture? Of the many definitions of culture, the one that seems most accurately to sum up this complex topic is 'a historical emergent set of values'. The cultural differences within the property/construction sectors may be seen to operate at a number of levels, but may be categorised as follows:

1. Business/economic factors (e.g. differences in the economic and legal systems, labour markets, professional institutions, etc. of different countries).
2. Anthropological factors, as explored by Hofstede (1984). The Hofstede IBM study involved 116,000 employees in forty different countries, and is widely accepted as being the benchmark study in this field.

Of these two groups of factors, the first may be regarded as fairly mechanistic in nature, and the learning curve for most organisations can be comparatively steep. For example, the practice of quantity surveyors in France of paying the contractor a sum of money in advance of any works on site may seem anathema, but it is usual practice in a system where the contractor is a trusted member of the project team. It is the second category of cultural factors, the anthropological factors, that is more problematic. This is particularly so for small and medium enterprises, as larger organisations have sufficient experience (albeit via a local subsidiary) to navigate a path through the cultural maze.

Perhaps one of the most famous pieces of research on the effects of culture was carried out by Gert Hofstede for IBM. Hofstede identified four key value dimensions on which national culture differed (Figure 5.3), a fifth being identified and added by Bond in 1988 (Hofstede and Bond, 1988). These value dimensions were power difference, uncertainty avoidance, individualism/collectivism and masculinity/femininity, plus the added long-/short-termism. Although neatly categorised and explained below, these values do of course in practice interweave and interact to varying degrees.

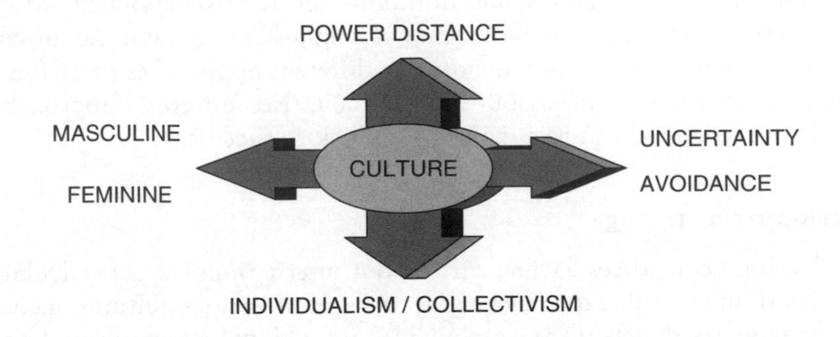

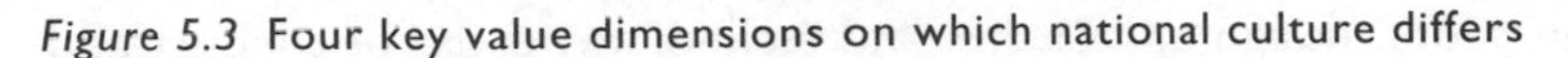
Figure 5.3 Four key value dimensions on which national culture differs

- *Power distance* indicates the extent to which a society accepts the unequal distribution of power in institutions and organisations, as characterised by organisations with high levels of hierarchy, supervisory control and centralised decision-making. For example, managers in Latin countries expect their position within the organisation to be revered and respected. For French managers the most important function is control, which is derived from hierarchy.
- *Uncertainty avoidance* refers to a society's ability to cope with unpredictability. Managers avoid taking risks and tend to play a higher role in planning and coordination. There is a tendency towards a greater quantity of written procedures and codes of conduct. In Germany, managers tend to be specialists and stay longer in one job, and feel uncomfortable with any divergence between written procedures (for example, the specification for concrete work and the works on site). They expect instructions to be carried out to the letter.
- *Individualism/collectivism* reflects the extent to which the members of a society prefer to take care of themselves and their immediate families as opposed to being dependent upon groups or other collectives. In these societies, decisions would be taken by groups rather than by individuals, and the role of the manager is as a facilitator of the team (e.g. Asian countries). In Japan tasks are assigned to groups rather than to individuals, creating stronger links between individuals and the company.
- *Masculinity/femininity* refers to the bias towards an assertive, competitive, materialistic society (masculine) or the feminine values of nurturing and relationships. Masculine cultures are characterised by a management style that reflects the importance of producing profits, whereas in a feminine culture the role of the manager is to safeguard the well-being of the workforce. To the American manager, a low head count is an essential part of business success and high profit; anyone thought to be surplus to requirements will be told to clear his or her desk and leave the company.

As a starting point for an organisation considering looking outside of the UK for work, Figure 5.4 may be a somewhat light-hearted but useful discussion aid to help recognise and identify the different approaches to be found towards organisational behaviour in other countries/cultures – approaches that if not recognised can be a major roadblock to success.

Developing a strategy

The development process, when carried out internationally, is particularly complex to manage due to the weaving together of various cultures, including language (both generic and technical), professional standards and construction codes, design approaches and technology, codes of conduct and

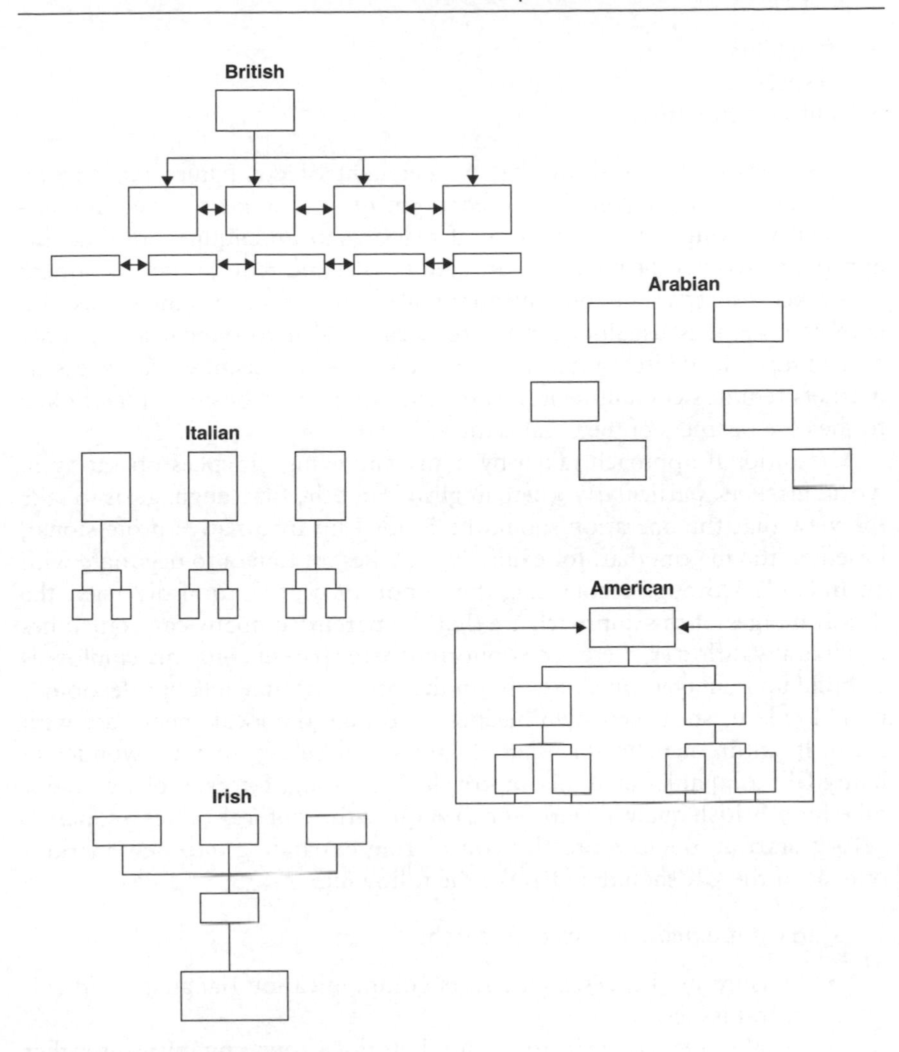

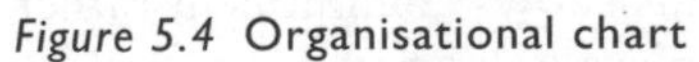

Figure 5.4 Organisational chart

Source: Adapted from International Management. Reed Business Publishing.

ethical standards. Technical competency and cultural integration must be taken as read.

The competencies necessary to achieve cultural fluency may therefore be said to be:

- interpersonal skills
- linguistic ability
- motivation to work abroad
- tolerance of uncertainty

- flexibility
- respect
- cultural empathy.

Case studies of SMEs show that 60 per cent of companies react to an approach from a company in another country to become involved in international working. The advantages of reacting to an enquiry are that this approach involves the minimum amount of risk and requires no investment in market research, but consequentially it never approaches the status of a core activity, it is usually confined to occasional involvement and is only ever of superficial interest. However, to be successful the move into overseas markets requires commitment, investment and a good business plan linked to the core business of the organisation.

A traditional approach taken by many surveying practices operating in world markets, particularly where English is not the first language, is to take the view that the operation should be headed up by a native professional, based on the maxim that, for example, 'it takes an Italian to negotiate with an Italian'. Although recognising the importance of cultural diversity, the disadvantages of this approach are that the parent company can sometimes feel like a wallflower, there is no opportunity for parent company employees to build up management skills, and in the course of time local professionals may decide to start their own business and take the local client base with them. If culture is defined as shared values and beliefs, then no wonder so many UK companies take this approach. How long, for example, would it take for a British quantity surveyor to acquire the cultural values of Spain?

As a starting point, a practice considering expanding into new markets outside of the UK should undertake the following:

1. Carry out extensive market research:

   - Ensure market research covers communication (language and cultural issues).
   - Make frequent visits to the market; this shows commitment rather than trying to pick up the occasional piece of work.
   - Use written language to explain issues, since verbal skills may be less apparent.
   - Use exhibitions to obtain local market intelligence and feedback.

2. Ensure documentation is culturally adapted and not literally translated:

   - Brochures should be fully translated into local language on advice from local contacts.
   - Publish new catalogues in the local culture.
   - Set up a website in the local language, with the web manager able to respond to any leads – after all, if a prospective client is expecting

a fast response, waiting for a translator to arrive is not the way to provide it.
- Adapt the titles of the services offered to match local perceptions.
- Emphasise added-value services.

3. Depending on the country or countries being targeted, operate as, for example, a European or Asian company, rather than a British company with a multilingual approach – think global, act local:

   - Arrange a comprehensive, multi-level programme of visits to the country.
   - Set up a local subsidiary company or local office or, failing that, set up a foreign desk inside the head office operating as if it is in the foreign country (e.g. keeping foreign hours, speaking foreign language).
   - Change the culture of the whole company at all levels from British to European, Asian, as relevant.
   - Recruit local agents who have been educated in the UK, so that they have a good understanding of UK culture too.

4. Implement a whole-company development strategy:

   - Language strategy should be an integral part of a company's overall strategy as a learning organisation.
   - Identify the few individuals who can learn languages quickly and build on this.
   - Create in-house language provision.
   - Set up short-term student placements in the UK for foreign students, via a sponsored scheme such as the EU Leonardo da Vinci programmes.
   - Target markets whose specialist language ability gives a competitive edge (e.g. China).

5. Subcontract the whole export process to a specialist company:

   - Hire a company to provide an export package of contacts, liaison, translation, language training and so on.

6. Pool resources with other companies:

   - Share language expertise and expenses with other companies.

7. In joint ventures, collaboration can be based upon:

   - Equity/operating joint ventures, in which a new entity is created to carry out a specific activity. Seen as a long-term commitment, the new entity has separate legal standing.
   - Contractual ventures, in which no separate entity is created and instead firms cooperate and share the risk and rewards in clearly

specified and predetermined ways. On the face of it, this form of joint venture appears to be more formal.

8. Management contracts:
   - The transfer of managerial skills and expertise in the operation of a business in return for remuneration.

Conclusion

With the advent of electronic communications, the possibilities that exist for quantity surveyors to operate on a European or global level have never been greater or easier to access. However, despite what some multinational organisations would have us believe, the world is not a bland, homogeneous mass and organisations still need to pay attention to the basics of how to conduct interpersonal relationships if they are to succeed.

> Ethical dilemmas solution:
>
> A) Accept the gift, pointing out that this does not ensure the awarding of future contracts.
>
> RICS members' confidential helpline is available to members on a 9 a.m. to 5 p.m. daily basis on +44 (0)20 7334 3867 and one of the range of matters on which members can get help and advice is ethical dilemmas.

Bibliography

Bardouil, S. (2001). Surveying takes on Europe. *Chartered Surveyor Monthly*, July/August, pp. 20–22.

Brooke, M.Z. (1996). *International Management*, 3rd edn. Stanley Thornes.

Button, R. and Mills, R. (2000). Public sector procurement: the Harmon case. *Chartered Surveyor Monthly,* 24 May.

Cartlidge, D. (1997). It's time to tackle cheating in EU public procurement. *Chartered Surveyor Monthly,* November/December, pp. 44–45.

Cartlidge, D. and Gray, C. (1996). *Cross Border Trading for Public Sector Building Work within the EU.* European Procurement Group, Robert Gordon University.

Cecchini, P. (1988). *The European Challenge.* Wildwood House.

CIOB (2006) *Corruption in the Construction Industry – A Survey.* The Chartered Institute of Building.

Commission of the European Communities (1985). *Completing the Internal Market.* Office for Official Publications of the European Communities.

Commission of the European Communities (1994). *Public Procurement in Europe: The Directives.* Office for Official Publications of the European Communities.

Commission of the European Communities (1998). *Single Market News – The Newsletter of the Internal Market DG.* Office for Official Publications of the European Communities.

Evans, R. (2009). Alan Wainwright: The lonely life of the construction industry whistleblower. *Guardian*, 15 May.

Fisher, C. and Lovell, A. (2009). *Business Ethics and Values, Third Edition.* Prentice Hall.

Hagen, S. (ed.) (1997). *Successful Cross-Cultural Communication Strategies in European Business.* Elucidate.

Hagen, S. (1998). *Business Communication Across Borders: A Study of the Language Use and Practice in European Companies.* The Centre for Information on Language Teaching and Research.

Hall, E.T. (1990). *Understanding Cultural Differences.* Intercultural Press.

Hall, M.A. and Jaggar, D.M. (1997). Should construction enterprises work internationally, take account of differences in culture? *Proceedings of the Thirteenth Annual ARCOM 97 Conference.* King's College, Cambridge, 15–17 September.

Hofstede, G. (1984). *Culture Consequences: International Differences in Work-Related Values.* Sage.

Hofstede, G. and Bond, M.H. (1988). *The Confuscius Connection: From Cultural Roots to Economic Growth.* Organizational Dynamics.

Matthews, D. (2010). Cover pricing 'as common now as it was two years ago'. *Building*, 11 June, p. 15.

McKendrick, P. (1998). RICS Annual Report and Accounts: President's Statement. *Chartered Surveyor Monthly*, 2 February.

McKillop, G. (2009). Fraud in construction – Follow the money. *Isurv*, December.

Moore, G. and Robson, A. (2002). The UK supermarket industry; an analysis of corporate and social responsibility. *Business Ethics; A European Review.*

Nash, L. (1981). Ethics without the sermon. *Harvard Business Review*, 59 (November/December).

Plimmer, F. (2009). *Professional Ethics – The European Code.* College of Estate Management, Reading.

RICS (1998). *Agenda for Change.* RICS.

RICS (2000). Professional ethics guidance note. RICS Professional Regulation and Consumer Protection Department.

RICS (2007a). *RICS Rules of Conduct for Firms.* RICS.

RICS (2007b). *RICS Rules of Conduct for Members.* RICS.

Tisser, M. *et al.* (1996). Chartered surveyors: an international future. *Chartered Surveyor Monthly,* 15 October.

Websites

http://simap.eu.int/
www.ted.eur-op.eu.int
www.tendersdirect.co.uk
www.cige.eu

Chapter 6

Delivering added value

Introduction

This chapter examines the ways in which quantity surveyors are delivering a range of added-value services to clients, based upon increased client focus and greater understanding of the function of built assets including such basic questions as: why are new buildings commissioned in the first instance and what is their function? Many construction clients who operate in highly competitive global markets are increasingly basing procurement strategies upon the degree of added value that can be demonstrated by a particular strategy. In order to meet these criteria, quantity surveyors must: 'get inside the head' of their clients, fully appreciate their business objectives and find new ways to delivered value, and thereby conversely remove waste from the procurement and construction process.

Procurement

The dictionary definition of procurement is 'the management of obtaining goods and services'. For many years quantity surveyors took this to mean appointing a contractor who submitted the lowest, that is to say the cheapest tender price, based upon a bill of quantities and drawings, in competition with several other contractors. For public sector works, a low-priced tender was almost guaranteed to win the contract, as public entities had to be seen to be spending public money prudently and would have had to develop a very strong case not to award a contract on the basis of the cheapest price. It is now clear that assembling an ad hoc list of six or so contractors, selected primarily on their availability, to tender for building work, for which they have very little detailed information, is not the best way to obtain value for money. In fact it does little more than reinforce the system in which contractors submit low initial prices secure in the knowledge that the contract will bring many opportunities to increase profit margins in the form of variation orders, claims for and extensions of time/loss and expense or mismanagement by the design team. Pre-Latham, the overriding ethos for procuring building works was to treat the supply chain with great suspicion; it was almost as if a cold war existed between client, design team and contractor/subcontractors. The emphasis in construction procurement has now swung away from the system described above towards systems that encourage partnerships and inclusion of the supply chain at an early stage – in fact to a point where the definition of procurement could be restated as 'obtaining value-for-money deals'. As discussed in Chapter 1, there can now be few individuals involved within the construction process who do not believe that the design, procurement and construction of new built assets has to become more efficient and client orientated. The evidence of wastage, in terms of materials, time and money not only in the short term, but also throughout the life cycle of a building, leaves the UK construction industry as well as its associated professions, including quantity surveyors, in an embarrassing positions and open to criticism from all sides for participating in the production of such a low-value product. For whatever reason, the quantity surveyor and the traditional brick-counting image enthusiastically fostered by so many within the industry, including the trade press, has also been the focus for this 'out-of-touch' image. For many years the quantity surveyor has been seen as the accountant to the construction industry, a knight in shining armour, safeguarding the client to ensure that they receive a building as close as possible to the initial agreed target price, although in practice this has been seldom achieved. Traditionally, a target cost has been set by the quantity surveyor, in discussion with the client at the outset, and then the process has been worked backwards, squeezing, in turn; the contractors, subcontractors and suppliers in order to keep within this target cost. The squeezing increases in direct proportion to the further down the supply chain the organisation comes (Figure 6.1).

The consequences of the supply chain squeeze illustrated in Figure 6.1 are low profit margins, lack of certainty and continuity for suppliers, delays in production, lack of consideration of whole-life costs, suboptimal functionality and rampant waste. In turn, low profits ensure that few, if any, resources can be channelled into research, technological improvement or quality assurance procedures. The Housing Grants, Construction and Regeneration Act 1996 did, to a certain extent, make the position of subcontractors more secure; however, their position in the supply chain still makes them vulnerable.

Construction productivity lags behind that of manufacturing and yet manufacturing has been a reference point and a source of innovation in construction for many decades; for example, industrialised building, currently undergoing a resurgence in interest in the UK, and the use of computer-aided design come directly from the manufacturing sector. However, while some innovations have crossed the divide from manufacturing to construction there has been little enthusiasm for other production philosophies. However, new manufacturing industry-based approaches to supply chain

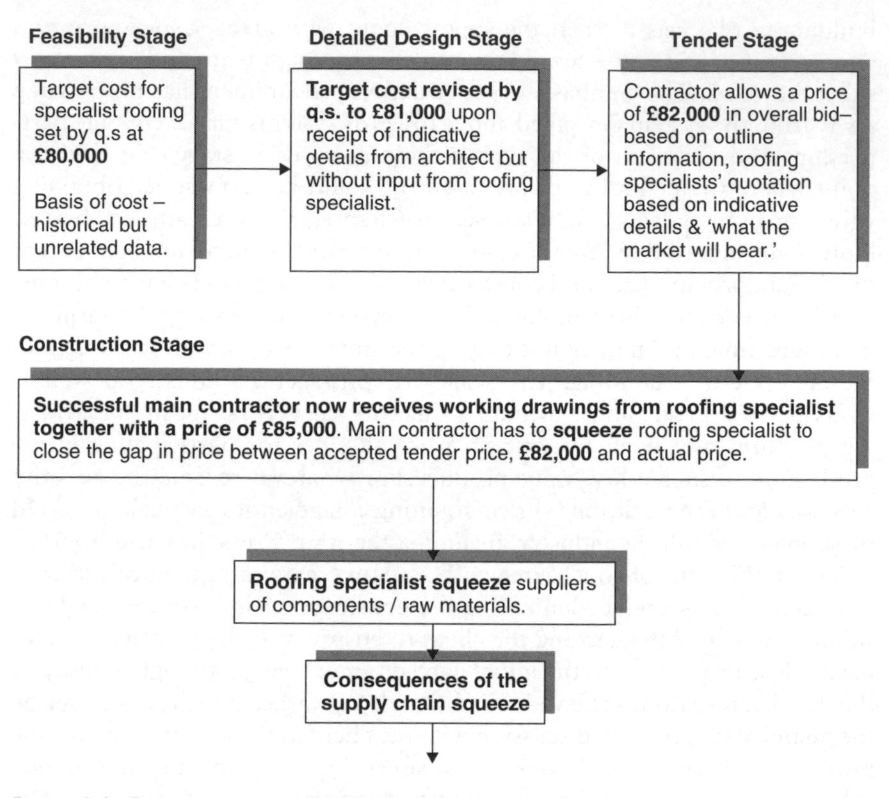

Figure 6.1 Consequences of the supply chain squeeze

management are now being used by the construction industry to enable the early involvement of suppliers and subcontractors in a project with devolved responsibility for design and production of a specific section of a building, with predicted and guaranteed whole-life costs, for periods of up to thirty-five years. The benefits of this new approach for clients include the delivery of increased functionality at reduced cost and, for the supply chain members: certainty, less waste and increased profits.

Supply chain relationships and management

A construction project organisation is usually a temporary organisation designed and assembled for the purpose of the particular project. It is made up of different companies and practices, which have not necessarily worked together before and which are tied to the project by means of varying contractual arrangements. This is what has been termed a temporary multi-organisation; its temporary nature extends to the workforce, which may be employed for a particular project rather than permanently. These traditional design team/supply chain models are the result of managerial policy aimed at sequential execution and letting out the various parts of the work at apparently the lowest costs. The problems for process control and improvement that this temporary multi-organisation approach produces are related to:

- Communicating data, knowledge and design solutions across the organisation.
- Stimulating and accumulating improvement in processes that cross the organisational borders.
- Achieving goal congruity across the organisation.
- Stimulating and accumulating improvement inside an organisation with a transient workforce.

The following quote, attributed to Sir Denys Hinton speaking at the RIBA seems to sum up the traditional attitude of building design teams:

> the so-called building team. As teams go, it really is rather peculiar, not at all like a cricket eleven, more that a scratch bunch consisting of one batsman, one goal keeper, a pole vaulter and a polo player. Normally brought together for a single enterprise, each member has different objectives, training and techniques and different rules. The relationship is unstable, and with very little functional cohesion and no loyalty to a common end beyond that of coming through unscathed.

Most of what is encompassed by the term 'supply chain management' was formerly referred to by other terms such as 'operations management' but the coining of a new term is more than just new management speak; it reflects the significant changes that have taken place across this sphere of activity. These changes result from changes in the business environment. Most manufacturing companies are only too aware of such changes: increasing globalisation, savage price competition, increased customer demand for enhanced quality and reliability, and so on. Supply chain management was introduced in order that manufacturing companies could increase their competitiveness in an increasingly global environment as well as their market share and profits by:

- Minimising the costs of production on a continuing basis.
- Introducing new technologies.
- Improving quality.
- Concentrating on what they do best.

The contrast between traditional approaches and supply chain management may be summed up as shown in Figure 6.2.

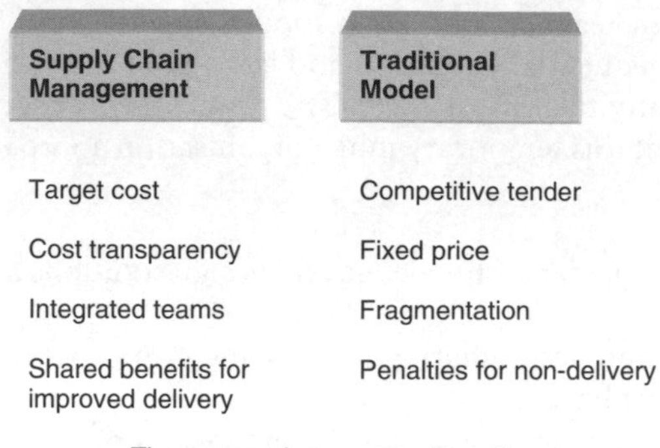

Figure 6.2 The contrast between traditional approaches and supply chain management

The quantity surveyor as supply chain manager

What is the driving force for the introduction of supply chain management into the UK construction industry? As described in Chapter 1, the CRINE initiative in the oil industry was the result of the collapse of world oil prices to what seems like today an unbelievable $13 a barrel in 1992; however, in construction very little impetus has come from the industry; it is clients who are the driving force. As discussed elsewhere, unlike other market sectors, because the majority of organisations working in construction are small, the industry has no single organisation to champion change. When Latham called for a 30 per cent reduction in costs, the knee-jerk response from some quarters of the profession and industry was that cost = prices and therefore it was impossible to reduce the prices entered in the bill of quantities by this amount; therefore the target was unrealistic and unachievable. However, this was not what Latham was calling for, as will be demonstrated in the following paragraphs. Reducing costs goes far beyond cutting the prices entered in the bill of quantities, if it ever did, it extends to the reorganisation of the whole construction supply chain in order to eliminate waste and add value. The immediate implications of supply chain management are:

- Key suppliers are chosen on criteria, rather than job by job on competitive quotes.
- Key suppliers are appointed on a long-term basis and proactively managed.
- All suppliers are expected to make sufficient profits to reinvest.

How many quantity surveyors have asked themselves the following question at the outset of a new project?

'What does value mean for my client?'

In other words, in the case of a new plant to manufacture, say, pharmaceutical products, what is the form of the built asset that will deliver value for money over the life cycle of the building for that particular client? For many years, whenever clients have voiced their concerns about the deficiencies in the finished product, all too often the patronising response from the profession has been to accuse the complainants of a lack of understanding in either design or the construction process, or both. The answer to the value question posed above will of course vary between clients; a large multinational manufacturing organisation will have a different view of value to a wealthy individual commissioning a new house, but it helps to illustrate the revolution in thinking and attitudes that must take place. In general, the definition of value for a client is 'design to meet a functional requirement for a through-life cost'. Quantity surveyors are increasingly developing better client focus, because only by knowing the ways in which a particular client perceives or even measures value, whether in a new factory or a new house, can the construction process ever hope to provide a product or service that matches these perceptions. Once these value criteria are acknowledged and understood quantity surveyors have a number of techniques, described in this chapter, at their disposal in order to deliver to their clients a high degree of the feel-good factor.

Not all of the techniques are new; many practising quantity surveyors would agree that the strength of the profession is expertise in measurement, and in supply chain management there is a lot to measure; for example:

- Measure productivity – for benchmarking purposes.
- Measure value – demonstrating added value.
- Measure out-turn performance – not the starting point.
- Measure supply chain development – are suppliers improving as expected?
- Measure ultimate customer satisfaction – customers at supermarkets, passengers at airport terminals, etc.

But of course, measuring value is not always easy to do.

What is a supply chain?

Before establishing a supply chain or supply chain network, it is crucial to understand fully the concepts behind and the possible components of a complete and integrated supply chain. The term 'supply chain' has become used to describe the sequence of processes and activities involved in the complete manufacturing and distribution cycle – this could include everything from product design through materials and component ordering through

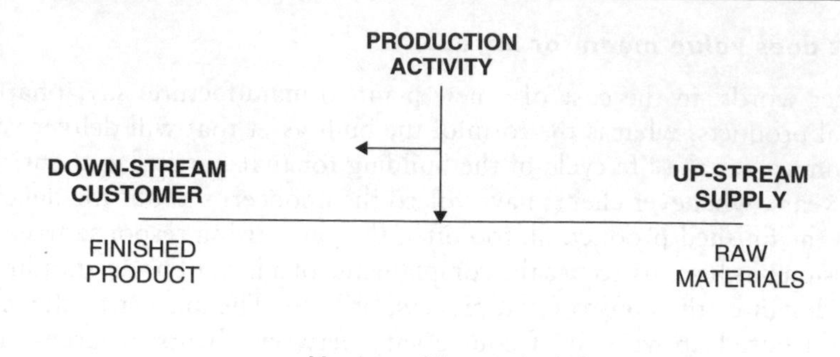

Figure 6.3 Members of the supply chain

manufacturing and assembly until the finished product is in the hands of the final owner. Of course, the nature of the supply chain varies from industry to industry. Members of the supply chain may be referred to as upstream and downstream supply chain members (Figure 6.3). Supply chain management, which has been practised widely for many years in the manufacturing sector, therefore refers to how any particular manufacturer involved in a supply chain manages its relationship both up and downstream with suppliers to deliver cheaper, faster and better. In addition, good management means creating a safe commercial environment in order that suppliers can share pricing and cost data with other supply team members.

The more efficient or lean the supply chain the more value is added to the finished product. As if to emphasise the value point, some managers substitute the word value for supply to create the value chain. In a construction context supply chain management involves looking beyond the building itself and into the process, components and materials which make up the building. Supply chain management can bring benefits to all involved, when applied to the total process which starts with a detailed definition of the client's business needs, which can be provided through the use of value management and ends with the delivery of a building which provides the environment in which those business needs can be carried out with maximum efficiency and minimum maintenance and operating costs. In the traditional methods of procurement the supply chain does not understand the underlying costs; hence suppliers are selected by cost and then squeezed to reduce price and whittle away profit margins.

- Bids based on designs to which suppliers have no input. No buildability.
- Low bids always won.
- Unsustainable – costs recovered by other means.
- Margins low, so no money to invest in development.
- Suppliers distant from final customer so took limited interest in quality.

The traditional construction project supply chain may be described as a series of sequential operations by groups of people or organisations (Figure 6.4).

Supply chains are unique, but it is possible to classify them generally by their stability or uncertainty on both the supply side and the demand side. On the supply side, low uncertainty refers to stable processes, while high uncertainty refers to processes which are rapidly changing or highly volatile. On the demand side, low uncertainty would relate to functional products in a mature phase of the production life cycle, while high uncertainty relates to innovative products. Once the chain has been catagorised, the most appropriate tools for improvement can be selected.

The construction supply chain is the network of organisations involved in the different processes and activities that produce the materials, components and services that come together to design, procure and deliver a building. Traditionally it is characterised by lack of management, little understanding between tiers of other tiers functions or processes and lack of communications, and a series of sequential operations by groups of people who have no concern about the other groups or the client. Figure 6.4 illustrates part of a typical construction supply chain, although in reality many more subcontractors could be involved. The problems for process control and improvement that the traditional supply chain approach produces are related to the following factors:

- The various organisations come together on a specific project, at the end of which they disband to form new supply chains.
- Communicating data, knowledge and design solutions across the organisations that make up the supply chain.
- Stimulating and accumulating improvement in processes that cross the organisational borders.

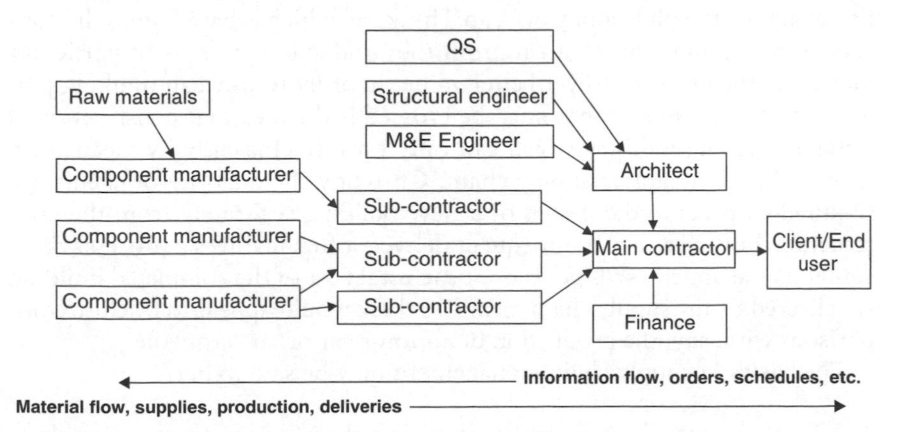

Figure 6.4 The traditional construction project supply chain

- Achieving goals and objectives across the supply chain.
- Stimulating and accumulating improvement inside an organisation that only exists for the duration of a project.

However, supply chain management takes a different approach that includes the following:

- Prices are developed and agreed, subject to an agreed maximum price, with overheads and profit being ring fenced. All parties collaborate to drive down cost and enhance value with, for example, the use of an incentive scheme.
- With cost determined and profit ring fenced, waste can now be attacked to bring down price and add value with an emphasis on continuous improvement.
- As suppliers account for 70 to 80 per cent of building costs they should be selected on their capability to deliver excellent work at competitive cost.
- Suppliers should be able to contribute new ideas, products and processes.
- Suppliers should be able to build alliances outside of projects.
- Suppliers should manage projects so that waste and inefficiency can be continuously identified and driven out.

The philosophy of integrated supply chain management is based upon defining and delivering client value through established supplier links that are constantly reviewing their operation in order to improve efficiency. There are now growing pressures to introduce these production philosophies into construction and it is quantity surveyors with their traditional skills of cost advise and project management who can be at the forefront of this new approach. For example, the philosophy of Lean Thinking, which is based upon the concept of the elimination of waste from the production cycle, is of particular interest in the drive to deliver better value. In order to use lean thinking philosophy the first hurdle that must be crossed is the idea that construction is a manufacturing industry which can only operate efficiently by means of a managed and integrated supply chain. Currently the majority of clients are required to procure the design of a new building separately from the construction; however, as the subsequent delivery often involves a process where sometimes as much as 90 per cent of the total cost of the completed building is delivered by the supply chain members there would appear to be close comparisons with, say, the production of a motor car or an aeroplane.

The basics of supply chain management may be said to be:

1. To determine which are the strategic suppliers, and concentrate on these key players as the partners who will maximise added value.

2. Work with these key players to improve their contribution to added value.
3. Designate these key suppliers as the 'first tier' on the supply chain and delegate to them the responsibility for the management of their own suppliers, the 'second tier' and beyond.

To give this a construction context, the responsibility for the design and execution of, say, mechanical installations could be given to a 'first-tier' engineering specialist. This specialist would in turn work with its 'second-tier' suppliers as well as with the design team to produce the finished installation. Timing is crucial, as first-tier partners must be able to proceed in the confidence that all other matters regarding the interface of the mechanical and engineering installation with the rest of the project have been resolved and that this element can proceed independently. However, at least one food retail organisation using supply chain management for the construction of its stores still places the emphasis on the tier partners to keep themselves up to date with progress on the other tiers, as any other approach would be incompatible with the rapid timescales that are demanded.

Despite the fact that on the face of it, certain aspects of the construction process appear to be prime candidates for this approach, the biggest obstacles to be overcome by the construction industry in adopting manufacturing industry-style supply chain management are as follows:

1. Unlike manufacturing, the planning, design and procurement of a building is at present separated from its construction or production.
2. The insistence that unlike an aeroplane or a motor-car, every building is bespoke, a prototype, and therefore is unsuited to this type of model or for that matter any other generic production sector management technique. This factor manifests itself by:
   - geographical separation of sites that causes breaks in the flow of production;
   - discontinuous demand;
   - working in the open air, exposed to the elements: can there be any other manufacturing process, apart from shipbuilding, that does this?
3. Reluctance by the design team to accept early input from suppliers and subcontractors, and unease with the blurring of traditional roles and responsibilities.

There is little doubt that the first and third hurdles are the result of the historical baggage outlined in Chapter 1 and that, given time, they can be overcome, whereas the second hurdle does seem to have some validity; despite statements from the proponents of production techniques buildings are not unique and commonality even between apparently differing building types is as high as 70 per cent (see Ministry of Defence, 1999). Interestingly, though, one of the main elements of supply chain management, Just in Time (JIT), was reported to have started in the Japanese shipbuilding industry in the mid-1960s, the very industry that opponents of JIT in construction quote as an example where, like construction, supply chain management techniques are inappropriate. Therefore, the point at which any discussion of the suitability of the application of supply chain management techniques to building has to start with the acceptance that construction is a manufacturing process, which can only operate efficiently by means of a managed and integrated supply chain. One fact is undeniable. At present, the majority of clients are required to procure the design of a new building separately from the construction. Until comparatively recently, international competition, which is a major influencing factor in manufacturing, was relatively sparse in domestic construction of major industrialised countries.

Adding value and minimising waste

The primary focus therefore in the design of new-build assets is on minimising value loss, whereas in construction it is on minimising waste.

The following statistics are taken from various reports from CIRIA and Movement for Innovation:

- Every year in the UK approximately thirteen million tons of construction materials are delivered to site and thrown away – unused.
- Ten per cent of products are wasted through oversupply, costing £2.4 billion per annum.
- Another £2.4 billion per annum is wasted in stockpiling materials.
- A further £5 billion per annum is squandered through the misuse of materials.

The lean construction toolbox

One of the most powerful and useful tools in the lean construction toolbox is value management.

Value analysis/engineering/management

Central to the goal of delivering built assets that meet the functional and operational needs of a client are the techniques of value engineering and value management. Developed first in the United States of America for the manufacturing and production sectors by Lawrence D. Miles, in the immediate post-Second World War era as value analysis, later relabelled as value engineering/management, this approach is now widely practised by UK quantity surveyors in both the public and private sectors. To quote

Robert N. Harvey, one-time manager of capital programmes and value management for the Port Authority of New York and New Jersey, 'Value Engineering is like love – until you've experienced it you just can't begin to understand it.' In the early 1990s the Port Authority conducted value engineering workshops on nearly $1 billion worth of construction projects. The total cost of the workshops was approximately $1 million, a massive statement of confidence in the technique that paid off, delivering nearly $55 million in potential savings.

For a somewhat more objective view of the process perhaps the reference point should be the International Society of American Value Engineers (SAVE), whose definition of value engineering is:

> A powerful problem-solving tool that can reduce costs while maintaining or improving performance and quality. It is a function-oriented, systematic team approach to providing value in a product or service.

The philosophy of value engineering/management is a step change from the traditional quantity surveying belief that delivering value is based upon the principle of cutting costs to keep within the original budget – what was and still is euphemistically referred to as cost reconciliation. Unlike this approach, the basis of value management is to analyse, at the outset, the function of a building, or even part of a building, as defined by the client or end-user; then, by the adoption of a structured and systematic approach, to seek alternatives and remove or substitute items that do not contribute to the efficient delivery of this function, thereby adding value. The golden rule of value engineering/management is that as a result of the value process the function(s) of the object of the study should be maintained and if possible enhanced, but never diminished or compromised (Figure 6.5).

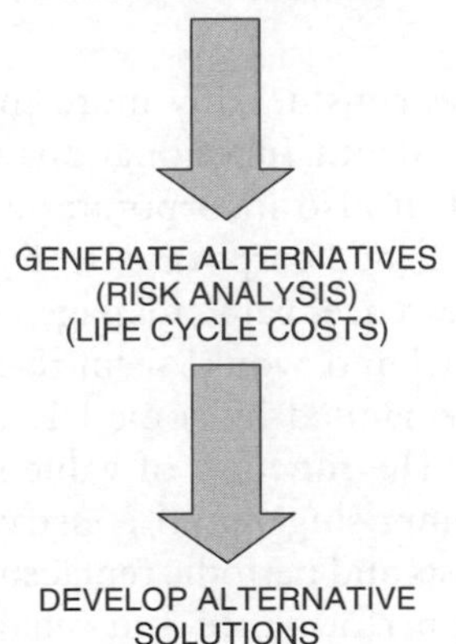

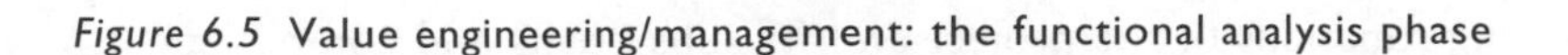

Figure 6.5 Value engineering/management: the functional analysis phase

Therefore, once again the focus for the production of the built asset is a client's perception of value. Perhaps before continuing much further the terms associated with various value methodologies should be explained. The terms in common usage are as follows.

Value analysis

The name adopted by Lawrence D. Miles for his early studies and defined as an organised approach to the identification and elimination of unnecessary cost.

Value engineering

The name adopted in 1959 by SAVE when it was established, to formalise the Miles approach. A term widely used in North America. The essential philosophy of VE is: 'A disciplined procedure directed towards the achievement of necessary function for minimum cost without detriment to quality, reliability, performance or delivery.' As if to emphasise the importance now being placed on value engineering, in 2000 Property Advisors to the Civil Estate (PACE) introduced an amendment to GC/Works/1 – Value Engineering Clause 40(4). The amendment states:

> The Contractor shall carry out value engineering appraisals throughout the design and the construction of the Works to identify the function of the relevant building components and to provide the necessary function reliability at the lowest possible costs. If the Contractor considers that a change in the Employer's Requirements could effect savings, the Contractor shall produce a value engineering report.

Value management

Value management involves considerably more emphasis on problem-solving as well as exploring in depth functional analysis and the relationship between function and cost. It also incorporates a broader appreciation of the connection between a client's corporate strategy and the strategic management of the project. In essence, value management is concerned with the 'what' rather than the 'how' and would seem to represent the more holistic approach now being demanded by some UK construction industry clients, i.e. to manage value. The function of value management is to reduce total through-life costs comprising initial construction, annual operating, maintenance and energy costs and periodic replacement costs, without affecting and indeed improving performance and reliability and other required design parameters. It is a function-oriented study and is accomplished by evaluating functions of the project and its subsystems and components to determine alternative means of accomplishing these functions at lower cost. Using value management, improved value may be derived in three predominant ways:

1. Providing for all required functions, but at a lower cost.
2. Providing enhanced functions at the same cost.
3. Providing improved function at a lower cost – the Holy Grail.

Among other techniques, value management uses a value engineering study or workshop that brings together a multidisciplinary team of people who are independent from the design team but who own the problem under scrutiny and have the expertise to identify and solve it. A value engineering study team works under the direction of a facilitator, who follows an established set of procedures; for example, the SAVE Value Methodology Standard (see Figure 6.6) to review the project, making sure the team understands the client's requirements and develops a cost-effective solution. Perhaps the key player in a VE study is the facilitator or value management practitioner, who must within a comparatively short time ensure that a group of people work effectively together. People like Alphonse Dell'Isola, the Washington, DC-based practitioner, who have risen to be an icon in value management circles, have helped SAVE to prove their claim that value management produces savings of 30 per cent of the estimated cost for constructing a project and that, for every pound invested in a VE study, including participants' time and implementation costs, £10 is saved. Certainly, organisations which have introduced VE into their existing procurement process, for example, previously publicly owned water companies, London Underground and so on, all report initial savings of around 10 to 20 per cent. In some respects, value management is no more than the application of the standard problem-solving approach to building design. If there is one characteristic which makes VM/VE distinctive it is the emphasis given to functional analysis.

The techniques that may be used to define and analyse function are as follows:

- value trees
- decision analysis matrix
- Functional Analysis System Technique Diagrams (FAST) (see Figure 6.7)
- criteria scoring.

Once the function of an item has been defined, the cost or worth can be calculated and the worth/cost ratio scrutinised to determine value for money. Value management may therefore be said to be a holistic approach to managing value that includes the use of value engineering techniques.

The process

The theory of value management is – buy function, don't buy product.

While it is not the purpose of this book to give a detailed description of every stage of a value engineering workshop it is worth spending some time to explain the process as well as to examine in more detail the functional analysis phase (see Figure 6.6). Value management has its roots in the manufacturing sector where it has been around for many years and there can be problems applying the approach to the construction of a new building. Nevertheless, valuable insights into the functions behind the need for a new building and what is required to fulfil these functions can flow from a value engineering workshop, even if it is not as lengthy and detailed as the traditional forty hours programme used in North America. If nothing else, it may be the only time in the planning and construction of a project when all the parties – client, end-user, architect, quantity surveyor – sit down together to discuss the project in detail.

For example, let us look at a hypothetical project where a value engineering workshop is used, the construction of a new clinic for the treatment of cancer. There are numerous variations and adaptations of not only the approach to conducting a value engineering workshop, but also the preparation of items such as the FAST diagram (see Figure 6.8). To illustrate the procedure, the following example is based on a classic forty-hour, five-day value engineering workshop, as this presents a more proven and pragmatic approach than some of the latter-day variations of value engineering. The workshop team is made up of six to eight experts from various design and construction disciplines who are not affiliated to the project, as it has been found that the process is not so vigorous if in-house personnel are used. In addition, an independent facilitator is recommended, as they have proved to be less liable to compromise on the delivery of any recommendations. The assembled team then commences the workshop, following the steps of the SAVE methodology (see Figure 6.7). At the start of the week the group is briefed on the project by the clinic personnel and members of the design and construction team, and the scope of the study is defined. Costs of the project are also carefully examined and analysed using a variety of techniques as well as compared to other facilities with a similar function. The first major task in the study is the functional analysis phase, during which the most beneficial areas for value improvement will be identified. While unnecessary cost removal has been the traditional target for quantity surveyors, it is important to emphasise that more frequently today value studies are conducted to improve a building performance without increasing cost, or, to express it more simplistically, to maximise 'bang for your buck'.

Functional analysis using a FAST model takes the following steps:

- Defining function.
- Classifying function.

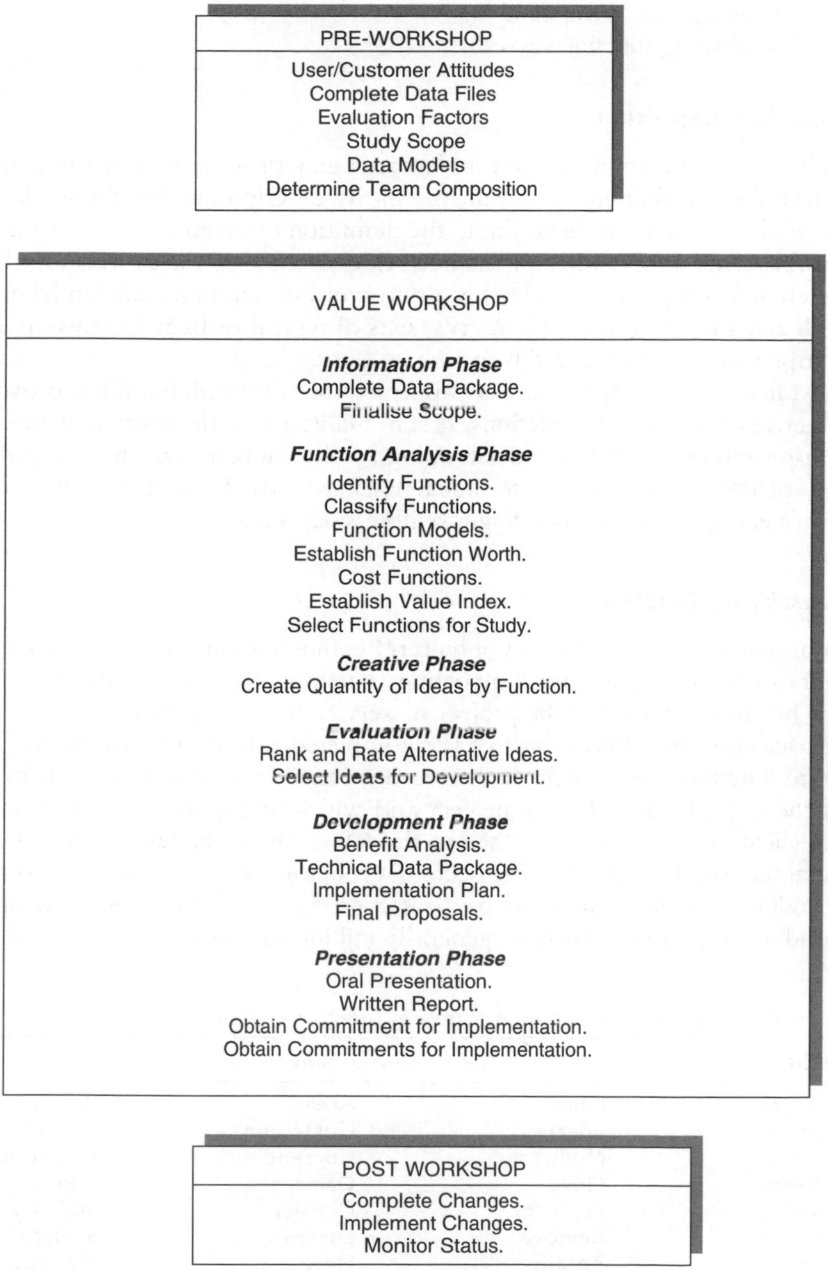

PRE-WORKSHOP

User/Customer Attitudes
Complete Data Files
Evaluation Factors
Study Scope
Data Models
Determine Team Composition

VALUE WORKSHOP

Information Phase
Complete Data Package.
Finalise Scope.

Function Analysis Phase
Identify Functions.
Classify Functions.
Function Models.
Establish Function Worth.
Cost Functions.
Establish Value Index.
Select Functions for Study.

Creative Phase
Create Quantity of Ideas by Function.

Evaluation Phase
Rank and Rate Alternative Ideas.
Select Ideas for Development.

Development Phase
Benefit Analysis.
Technical Data Package.
Implementation Plan.
Final Proposals.

Presentation Phase
Oral Presentation.
Written Report.
Obtain Commitment for Implementation.
Obtain Commitments for Implementation.

POST WORKSHOP

Complete Changes.
Implement Changes.
Monitor Status.

Figure 6.6 FAST Diagram

Source: Society of American Value Engineers.

- Developing function relationships.
- Assigning cost to function.
- Establishing function's worth.

Function: definition

Definition of function can be problematic; experience has shown that the search for a definition can result in lengthy descriptions that do not lend themselves to analysis. In addition, the definition of function has to be measurable. Therefore, a method has been devised to keep the expression of a function as simple as possible; it is a two-word description made up from a verb and a noun. Table 6.1 gives two lists of typical verbs and nouns; more comprehensive lists are readily available.

At first sight this approach may appear to be contrived, but it has proved effective in pinpointing functions. It is not cluttered with superfluous information and promotes full understanding by all members of the team regardless of their knowledge or technical backgrounds: for example, identify treatment, assess condition, diagnose illness (see Figure 6.8).

Classifying function

In order to establish some sort of hierarchy, functions are classified into primary or basic function and supporting functions. Basic functions or needs are functions that make the project or service work; if omitted this would impact upon the effectiveness of the completed project. Out of the list of basic functions emerges the highest order function, which may be defined as the overall reason for the project and which meets the overall needs of the client. This function is placed to the left of the scope line on the FAST diagram (see Figure 6.8). The second grouping of functions, supporting functions, may in the majority of cases contribute nothing to the value of a building. Supporting functions generally fall into the following categories:

Table 6.1 Typical verbs and nouns used in functional analysis

Verbs		*Nouns*	
Amplify	Limit	Area	Power
Attract	Locate	Corrosion	Protection
Change	Modulate	Current	Radiation
Collect	Move	Damage	Repair
Conduct	Protect	Density	Stability
Contain	Remove	Energy	Surface
Control	Rotate	Flow	Vibration
Enclose	Secure	Fluid	Voltage
Filter	Shield	Heat	Volume
Hold	Support	Insulation	Weight

- Assure dependability.
- Assure convenience.
- Satisfy user.
- Create image.

At first glance these categories may seem to have little relevance to construction-related activities until it is understood that, for example, the 'create image' heading includes items such as aesthetic aspects, overall appearance, decoration and implied performance, such as reliability, safety, etc.; items that in themselves are not vital for the integrity of the project, but nevertheless may be high on the client's list of priorities.

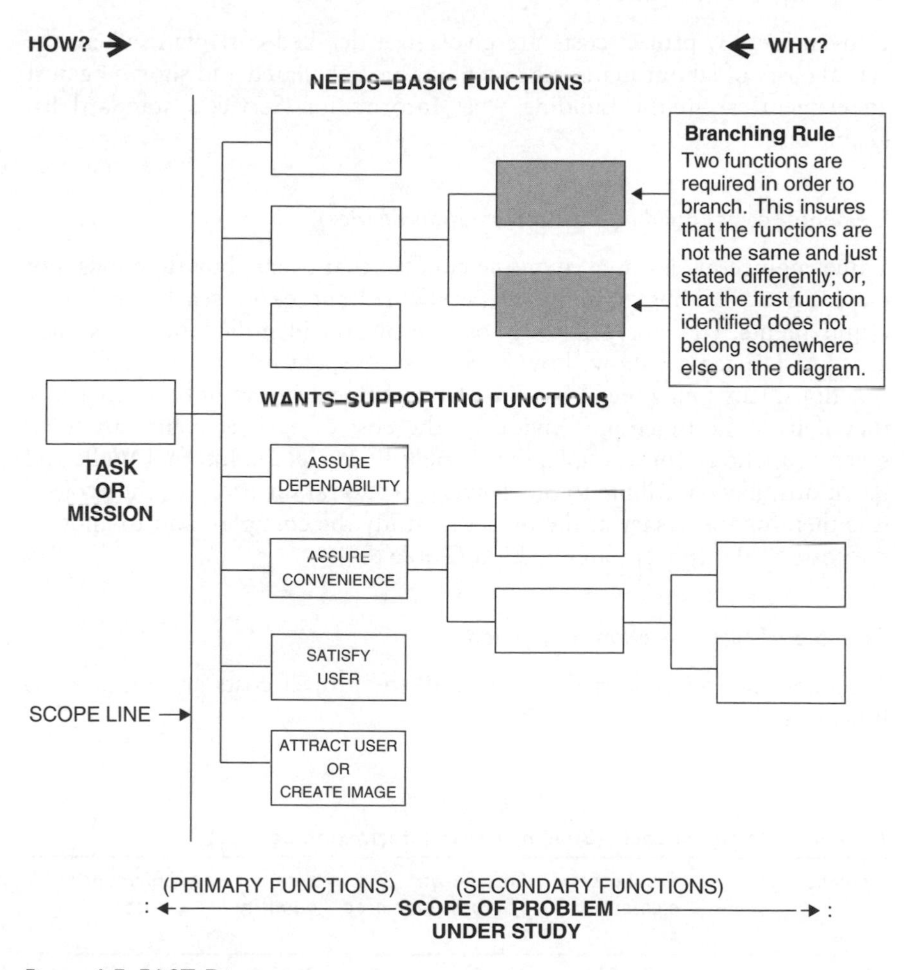

Figure 6.7 FAST Diagram

Source: SAVE International.

Developing functional relationships

Functional Analysis System Technique (FAST) models are a method of depicting functional relationships (see Figures 6.7 and 6.8). The model works both vertically and horizontally by first determining the highest order function, called the task or mission, that is positioned to the left of the vertical scope line. By working from the left and asking the question *How?* and employing the verb/noun combination and working from the right asking the question *Why?*, the functions and their interrelationship can be mapped and their value allocated at a later phase.

Assigning cost to function

Conventionally, project costs are given in a detailed cost plan, where the actual costs of labour materials and plant are calculated and shown against an element, as in the Building Cost Information Service's standard list (Table 6.2).

Elemental costs (Building Cost Information Service)

Value engineering is based upon the concept that clients buy functions, not materials or building components, as defined and expressed by their user requirements. Therefore, splitting costs among the identified functions, such as a FAST diagram, shows how resources are spent in order to fulfil these functions. Costs may then be viewed from the perspective of how efficiently they deliver the function. Obviously, the cost of each element can cover several functions; for example, the element BCIS Ref 2G Internal Walls and Partitions may contribute to the delivery of several functions of the project. It is therefore necessary at the outset to study the cost plan and to allocate the costs to the appropriate function (Table 6.3).

Example of cost allocation to function

A similar exercise is carried out until all the project costs are allocated to functions.

Table 6.2 Elemental costs (Building Costs Information Service)

Element	*Total cost of element* £	*Cost per m^2 of gross floor area* £	*Element unit quantity*	*Element unit rate* £
2G Internal walls and partitions	430,283	45.00	8025m^2	53.62

Table 6.3 Example of cost allocation to function

Element: 2G internal walls and partitions *Function*	*Elemental cost : £430,283* *Cost £*
3.1 INDENTIFY PATIENTS	12,000
3.2 MAINTAIN RECORDS	30,000
3.3 STUDY DIAGNOSIS	25,900
3.4 INCREASE AVAILABILITY	6,900
3.5 MAINTAIN HYGIENE	7,000
4.1 REFER PATIENT	45,000
4.2 TREAT PATIENT	56,000
4.3 PROCESS RECORDS	40,000
4.4 CIRCULATE PEOPLE	60,889
5.1 COUNSEL PATIENT	26,605
5.2 REDUCE STRESS	4,989
5.3 PROTECT PRIVACY	38,000
6.1 COMFORT PATIENT	34,000
6.2 APPEAR PROFESSIONAL	43,000
	£430,283

Establishing function's worth

The next step is to identify which of the functions contains a value mismatch, or in other words seems to have a high contribution to the total project cost in relation to the function that it performs. Following on from this the creative phase will concentrate on these functions. Worth is defined as 'the lowest overall cost to perform a function without regard to criteria or codes'. Having established the worth and the cost, the value index can be calculated. The formula is Value = Worth/Cost. The benchmark would be to achieve a ratio of 1.

The FAST diagram illustrated in Figure 6.8 is characterised by the following:

- The vertical 'scope line' which separates and identifies the highest level function – the task or mission – from the basic and supporting functions. It is pivotal to the success of a functional analysis diagram that this definition accurately reflects the mission of the project.
- The division of functions into needs or basic functions – with these functions the project will not meet client requirements and wants or supporting functions; these are usually divided into the four groups as previously discussed. The project could still meet the client's functional requirements if these wants are not met or included.
- The use of verb/noun combinations to describe functions.
- Reading the diagram from the left and asking the question, How is the function fulfilled? provides the solution.

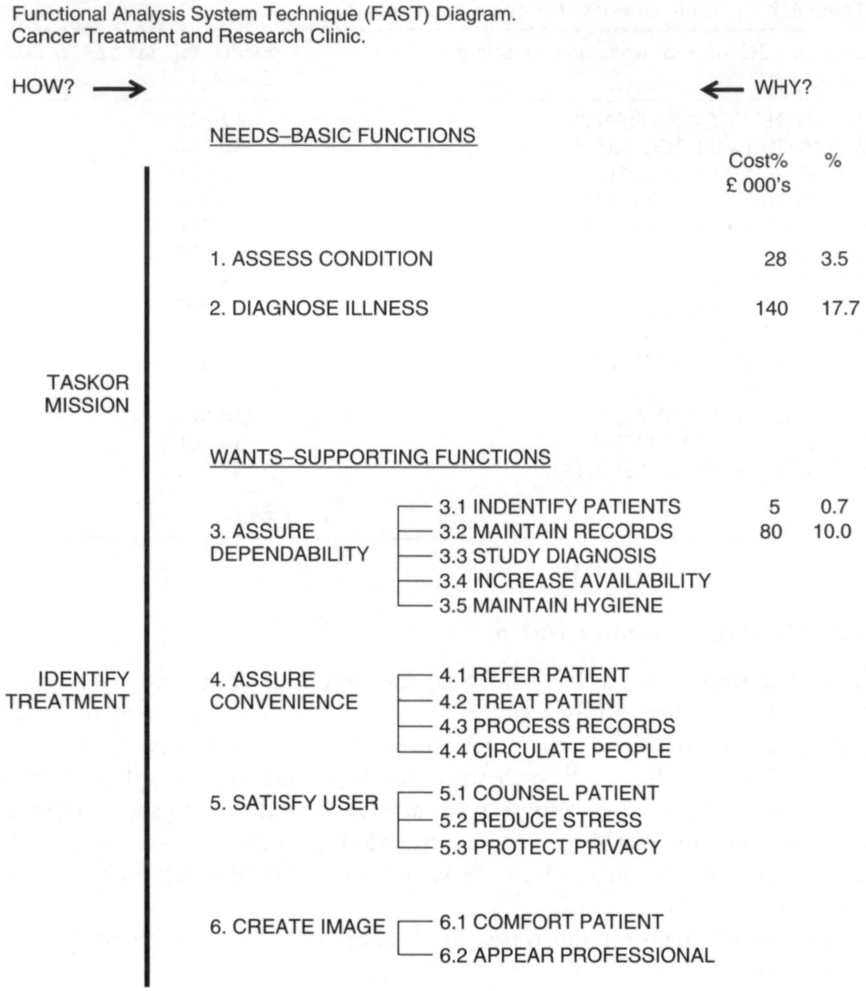

Functional Analysis System Technique (FAST) Diagram.
Cancer Treatment and Research Clinic.

HOW? → ← WHY?

NEEDS–BASIC FUNCTIONS

		Cost% £ 000's	%
	1. ASSESS CONDITION	28	3.5
	2. DIAGNOSE ILLNESS	140	17.7

TASKOR MISSION

WANTS–SUPPORTING FUNCTIONS

			Cost% £ 000's	%
IDENTIFY TREATMENT	3. ASSURE DEPENDABILITY	3.1 INDENTIFY PATIENTS	5	0.7
		3.2 MAINTAIN RECORDS	80	10.0
		3.3 STUDY DIAGNOSIS		
		3.4 INCREASE AVAILABILITY		
		3.5 MAINTAIN HYGIENE		
	4. ASSURE CONVENIENCE	4.1 REFER PATIENT		
		4.2 TREAT PATIENT		
		4.3 PROCESS RECORDS		
		4.4 CIRCULATE PEOPLE		
	5. SATISFY USER	5.1 COUNSEL PATIENT		
		5.2 REDUCE STRESS		
		5.3 PROTECT PRIVACY		
	6. CREATE IMAGE	6.1 COMFORT PATIENT		
		6.2 APPEAR PROFESSIONAL		

Figure 6.8 FAST diagram: Cancer Treatment and Research Clinic

- Reading the diagram from the right and asking the question 'Why?' identifies the need for a particular function.
- The right-hand side of the diagram allows the opportunity to allocate the cost of fulfilling the functions in terms of cost and percentage of total cost.

Therefore, the FAST diagram in Figure 6.8 clearly shows the required identified functions of the project, together with the cost of providing those functions. What now follows is the meat of the workshop – a creative session that relies on good classic brainstorming of ideas, a process that has been

PROJECT Cancer Treatment and Research Clinic	VALUE ENGINEERING PROPOSAL
PROPOSAL Eliminate return duct to ventilation system	DATE
	ITEM No H14

ORIGINAL PROPOSAL:

Each room has a return grille and ductwork connecting back to a return fan.

PROPOSED CHANGE:

Eliminate duct return system on individual floors and provide an above ceiling return plenum.

ADVANTAGES:

More available ceiling space

Balancing of return system is simplified

DISADVANTAGES:

Plenum rated cable, tubing and pipe required

May be acoustic transmission problems in walls

COST SUMMARY	INITIAL COST	OPERATION & MAINTENANCE COST–15 Years		TOTAL LIFE CYCLE COST
		PER ANNUM	LIFECYCLE–PV@ 6%	
ORIGINAL PROPOSAL	£149,450	£4,000	£38,848	£188,298
VE PROPOSAL	£86,000	£2,000	£19,424	£105,424

Figure 6.9 Value engineering outcome report

compared by those who have experienced it to a group encounter session, the aim of which is to seek alternatives. The discussion may be structured or unstructured – Larry Miles was quoted as saying that the best atmosphere to conduct a study was one laced with cigarette smoke and Bourbon, but in these more politically correct times such aids to creativity are seldom employed. The rules are simple. Nobody is allowed to say, 'That won't work'. Anybody can come up with a crazy idea. These sessions can generate hundreds of ideas, of which perhaps fifty will be studied further in the workshop's evaluation phase. Those ideas will be revisited and some discussion will take place as to their practicality and value to the client. Every project will have a different agenda. The best of the recommendations are then fully developed by the team, typically on day four of the workshop, and studies are carried out into costs and through-life costs of a proposed change before presentation to the client on the final day. It is an unfortunate fact of life of the classic five-day workshop that the team member tasked with costing the recommendations has to work into the night on the penultimate day. Finally, a draft report is approved and a final report is written by the team leader. In addition to the above procedures, risk assessment can or, as is thought in some circles, should be introduced into the process. As the value analysts go through and develop value recommendations they may be asked to identify risks associated with those recommendations, which can either be quantitative or qualitative. If brainstorming sounds just a little esoteric to the quantity surveying psyche, take heart: a value engineering workshop usually produces tangible results that clearly set out the costs and recommendations in a very precise format (see Figure 6.9).

The question is often asked: are there projects that are beyond value management? The answer is most certainly 'Yes'. There are many high-profile examples that flaunt the drive to lean construction and these mainly fall into the category of projects for which making a statement either commercially, politically or otherwise is their primary highest order function. Flyvbjerg, in *Megaprojects and Risk: An Anatomy of Ambition* (Flyvbjerg *et al.*, 2003), cites several examples of international mega-projects that have developed their own unstoppable momentum.

Managing the supply chain

Partnering and alliancing

The preferred approach to managing the supply chain is partnering – it has been described as welding the links of the supply chain together. Although the term 'partnering' is relatively new, having been adopted in various guises within the UK construction industry since the late 1980s, this is not the case with the relationship itself. Some contractors had been practising what they might term 'collaborative contracting' for many years before the term partnering was adopted with respect to a formal arrangement – for example, Bovis' relationship with Marks & Spencer.

Essentially, partnering enables organisations to develop collaborative relationships either for one-off projects (project-specific) or as long-term associations (strategic partnering). The process is used as a tool to improve performance, and may apply to two organisations (e.g. a client and a design-and-build contractor) or to a number of organisations within a formal or informal agreement (e.g. consultants, contractors, subcontractors, suppliers, manufacturers, either with or without client participation). The partnering process is formalised within a relationship that might be defined within a charter or a contractual agreement. Partnering is seen by many as a means of avoiding risks and conflict. There isn't one model partnering arrangement; it is an approach that is essentially flexible, and needs to be tailored to suit specific circumstances. In addition to partnering another collaborative approach to project delivery is alliancing, described later in this chapter. The term 'partnering' rather than 'partnerships' was chosen for this procurement strategy owing to the legal implications of partnerships. However, to some observers partnering is still an ambiguous term to which at least half a dozen different perspectives may be applied. Overall, partnering has received a mixed press in terms of improved performance. For instance, to some observers, partnering has been seen to try to impose a culture of win/win over the top of a commercial framework which remains inherently win/lose. Studies conducted across other industries where partnering and supply chain management is common suggest that despite the best intentions, clients easily revert to cost-based criteria rather than value for money, and that rarely do the supply chain members share a common purpose.

By having a smaller number of firms to work with, the client gains considerable benefits. The partnering organisations may gain greater experience of the client's needs, use techniques such as standardisation of components and processes and bulk purchasing, and achieve continuous improvement. Sharing lessons between organisations and applying new ideas ensures the client is getting best practice.

There is evidence, according to the Civil Engineering Contractors Association, that in the civil engineering industry it seems that some clients have entered into partnering arrangements with contractors, or have let framework agreements without fully appreciating all that is required for these arrangements to be successful in terms of delivering better value. In particular, some clients seem not to have looked much beyond the subsidiary objective of these arrangements, which is to secure savings in costs of procurement and contract administration. There is evidence of what has been labelled 'institutional pressure'; that is to say clients and contractors feeling that they must move in the directions in which the Latham and Egan reports are pointing them, but there is a danger that they will begin to move on the basis of insufficient knowledge and understanding of what is required. According to Wood, numerous authors have tried to analyse the critical success factors for successful partnering relationships; however, despite some differences from various studies, the assertion made by Bennett and Jayes that true partnering relies on cooperation and teamwork, openness and honesty, trust, equity and equality is still appropriate.

Overview – a client's perspective

Government-led initiatives have repeatedly expressed the wish to see partnering become the norm in the hope that it will promote a new way of working. It is clear that public sector clients (including local authorities) are being directed towards procurement strategies that are based upon integration and collaboration. The National Audit Office's report, *Modernising Construction*, gave support for public sector clients in promoting innovation and good practice, encouraging the industry as a whole and its clients to:

- Select contractors on the basis of value for money.
- Develop close working relationships among clients and the entire supply chain.
- Integrate the entire supply chain, including clients, professional advisers, designers, contractors, subcontractors and suppliers.

In the private sector, many major and influential clients across the board have been adopting partnering in response to the proven long-term benefits that may be achieved through this approach. Companies such as Sainsbury's, BAA and Esso are reported to have reached savings of up to 40 per cent on costs and 70 per cent on time by using partnering approaches. There is, however, scant evidence that small, occasional clients have little to gain from the process.

Overview – a contractor's and consultant's perspective

Main contractors, publicly at least, appear to have enthusiastically embraced the partnering concept, without which much of the work available from the major clients is not accessible. General and specialist subcontractors, suppliers and manufacturers may be involved through partnering within a larger supply chain, but many claim that the only benefit to them is assured workload, although this comes at a price – lower profit margins, for example. Nevertheless, it is interesting to note the growth of networking events/marketplaces that offer manufacturers the opportunity to forge new relationships by providing a consultation service on their stands to promote their design and problem-solving skills, rather than selling products. These collaborative, solutions-driven events have been enthusiastically supported by major materials manufacturers, who recognise the contribution they are already making to design through early involvement in partnering arrangements. Many large client organisations now have framework agreements with consultants as well as with contractors, covering periods of time and a series of projects, and of course many consultants have been 'preferred' firms of regular clients for many years. Framework agreements often do little more than formalise long-term relationships, although some clients are becoming more demanding of their consultants in these agreements, resulting in firms being dropped and others refusing to sign up. The main attraction for consultants, large or small, is undoubtedly the security of workload offered by long-term arrangements, but this may be at a price – financial or otherwise. Many consultants, particularly architects, have formed strong relationships with contractors to compete for design-and-build projects, which have been increasingly attracting many clients for some years. Some of these arrangements are now developing into the core of prime contracting alliances which are discussed later in this chapter. Besides security of workload, attractions of partnering for consultants might include the satisfaction and reputation gained from being associated with successful projects or high-profile clients. Theoretically, greater profits are achievable through sharing in savings; however, there is a lack of hard evidence that consultants benefit from this. Nevertheless, it is likely that, where consultants partner with contractors in, say, a design-and-build or prime contracting project, the consultants may well share some of the savings awarded to the contractor. Project-specific partnering would not appear to offer many benefits for consultants.

A RICS research report; *Beyond Partnering: Towards a New Approach in Project Management*, published in 2005, aimed to examine how the barriers to partnering success, culture and the economic reality of supply chain relationships are being addressed in practice. In addition, the study attempted to assess the actual benefits that accrue from partnering and whether there is any real change in construction industry practice. The RICS study used a series of semi-structured interviews with a relatively small sample of senior figures within ten major construction clients, including large retailers and utilities organisations with a combined annual construction spend of approximately £2,000 million. The client sample procured between 50 and 100 per cent of their total expenditure using partnering arrangements. In addition, ten national contracting organisations were selected with a combined annual turnover of approximately £4,350 million of which £1,500 million is delivered through partnering arrangements.

Respondents were unanimous that to be successful, partnering arrangements, which in some cases included alliancing (discussed below) require a culture change within the industry on both supply and demand sides. An inherent lack of trust manifested itself as a threat to successful partnering, which was identified as lack of openness and honesty of clients and a 'Luddite' culture within contractors' organisations, resulting in little change of practice at site level. Interestingly, the survey heard 'that it is quantity surveyors who find it most difficult to adapt, since they are used to problems being addressed in a contractual and confrontational manner, rather than by people communicating in order to find solutions'. On economic issues clients admit to obtaining commercial leverage over their supply chains and there is evidence that the subcontractor squeeze described previously in this chapter is still alive and well in supply chains. For contractors the continuity of working repeatedly for the same clients is also thought to provide a number of benefits for contracting organizations, although whether a contractor's profit margin increases on partnering projects is unclear. Both supply and demand sides agreed that partnering provided a more rewarding environment in which to operate.

Key success factors

Simply adopting a policy of partnering with and within the supply chain will not itself ensure success. Partnering is not an easy option; a number of prerequisites, or key success factors, need to be taken on board. Some of the following are desirable for project partnering; all are essential for successful strategic partnering:

- There needs to be a commitment at all levels within an organisation to make the project or programme of work a success.
- Partners must have confidence in each other's organisations, and each organisation needs to have confidence in its own team.

Clients should normally select their partners from competitive bids based on carefully set criteria aimed at getting best value for money. This initial competition should have an open and known prequalification system for bidders.

- Partners need collectively to agree the objectives of the arrangement/project/programme of work and ensure alignment/compatibility of goals.
- To satisfy the relationship's agenda, there needs to be clarity from the client and continued client involvement.
- Sharing is important. All players should share in success in line with their contribution to the value-added process (which will often be difficult to assess). There also needs to be a sharing of information, which requires open-book accounting and open, flexible communication between organisations/teams/people.
- It is important that all partnering arrangements incorporate effective methods of measuring performance. It has been identified that partnering should strive for continuous improvement, and this must be measurable to ascertain whether or not the process is effective.
- There will be times when partners don't agree, and it is therefore important that agreed non-adversarial conflict resolution procedures are in place to resolve problems within the relationship. The principle of

trying to resolve disputes at the lowest possible level should normally be adopted to save time and cost.
- Education and training is needed to ensure an understanding of partnering philosophy.

Opportunities for quantity surveyors

Collaborative integrated procurement offers opportunities for quantity surveyors, including:

- Acting as an independent client adviser. Many clients will still look to their quantity surveyor for independent advice. This raises the question, 'Where is the trust within the relationship if external advice is still needed?' Many clients will still feel that they need the advice of someone without an axe to grind – for example, appointing an external quantity surveyor and audit team to ensure that its strategic partners perform. Services provided might include assessment of target costs, development of incentive schemes, measurement of performance, auditing and so on.
- Participating as a partner in an alliance. Quantity surveyors able to demonstrate that they have the skills and ingenuity to add value will be welcomed by most alliances. Imaginative teams will consider numerous solutions – who better to evaluate the alternatives than the quantity surveyor?
- Leading an integrated supply chain. Many quantity surveyors have become successful project managers. There is no reason to suggest therefore that they cannot manage a supply chain. With appropriate financial resources, a quantity surveying practice can act as a prime contractor.
- Acting as a partnering adviser within PPC 2000 contracts. The described role would seem to fit the quantity surveyor with partnering experience. This is a key role, and would suit a quantity surveyor who can demonstrate a collaborative rather than adversarial attitude.
- The move towards clients partnering with integrated supply chains offers significant opportunities for consultants wishing to join alliances to share in the potential rewards. If the industry does become less adversarial, as is hoped, quantity surveyors will welcome it. They will then be able to concentrate on what they do best – adding value for clients, which coincides with the purpose of partnering!

Alliancing

As with partnering, alliancing may be catagorised as follows:

> Strategic alliances may be described as two or more firms that collaborate to pursue mutually compatible goals that would be difficult to

achieve alone. The firms remain independent following the formation of the alliance. Alliancing should not be confused with mergers or acquisitions.

A project alliance is where a client forms an alliance with one or more service providers; designers, contractors, supplier, etc. for a specific project and this section will continue to concentrate on this aspect of alliancing.

The principal features of a project alliance are as follows:

- The project is governed by a project alliance board, which comprises all parties to the alliance that have equal representation on the board. One outcome of this is that the client has to divulge to the other board members far more information than would, under other forms of procurement, be deemed to be prudent.
- The day-to-day management of the project is handled by an integrated project management team drawn from the expertise within the various parties on the basis of the best person for the job.
- There is a commitment to settle disputes without recourse to litigation except in the circumstance of wilful default.
- Reimbursement to the non-client parties is by way of 100 per cent open-book accounting.

A fundamental principle of alliances is the acceptance on the part of all the members of a share of losses, should they arise, as well as a share in rewards of the project. Risk:Reward should be linked to project outcomes which add to or detract from the value to the client. In practice there will be a limit to the losses that any of the alliance members, other than the client, will be willing to accept, if the project turns out badly. Unless there are good reasons to the contrary it may be expected that the alliance will take 50 per cent of the risk and the owner/client the remaining 50 per cent.

The major differences between alliancing and project partnering are set out in Table 6.4.

In project partnering one supplier may sink or swim without necessarily affecting the business position of the other suppliers. One entity may make a profit, while the other makes a financial loss. However, with alliancing there is a joint rather than a shared loss; therefore, if one alliance party underperforms, all the parties are at risk of losing.

Partnering contracts

Since the 1990s specific contracts and contractual amendments have been drafted to implement partnering as part of the formal contract process. These include:

Table 6.4 Major differences between alliancing and project partnering

	Partnering	*Alliances*
The form of the undertaking	Core group with no legal responsibilities. Binding/non-binding charters used in 65% of partnering arrangements.	Quasi joint venture operating at one level as a single company.
The selection process	Prime contractor responsible for choice of supply chain partners.	Rigorous selection process.
	Project can commence while selection continues.	Alliance agreement not concluded until all members appointed.
The management structure	By prime contractor. Partnering adviser.	Alliance board.
Risk-and-reward mechanisms	Partners' losses not shared by other members of the supply chain.	Losses by one alliance member shared by other members.

- The ACA Standard Form of Contract for Project Partnering (PPC2000).
- The New Engineering Contract 3rd Edition (NEC3) has Option X12 set of clauses; this in effect is a Standard Form of Contract underneath a partnering agreement.

In addition, the following two alternatives are available:

- No contract but instead a partnering charter.
- Standard form of contract (JCT 05) and partnering charter.

Project Partnering Contract 2000 (PPC2000)

Even though partnering has been described as a state of mind rather than the opportunity to publish another new contract, there is now available a standard form of contract for use with partnering arrangements – Project Partnering Contract 2000 (PPC 2000). PPC 2000 was developed for individual projects that were to be procured using the partnering ethos. By the nature of the contract it is more suitable to a single project rather than an ongoing framework agreement covering many projects over a period of time, which would need a bespoke contract. PPC 2000 was developed to clarify the contractual relationship where partnerships were being entered into using amended JCT contracts, and in many cases this was proving to be confusing and to the JCT losing its familiarity. Partnering charters, while proving valuable, were not a legal basis for an agreement and did not outline the system for non-adversarial approach. Drafted by Trowers and Hamlins, a leading London law firm, PPC was launched in November 2000 by Sir John Egan and published by the Association of Consultant Architects Ltd (ACA). There is a standard adaptation for use in Scotland.

The multi-party PPC 2000 contract

The aim of PPC 2000 is that the team members are in contract much earlier than is traditional and for negotiations to be carried out post-contract. The contract still allows for a partnering charter and also covers consultants' appointments and agreements – one contract for the whole. It is claimed that PPC 2000 is written in straightforward language with the flexibility to allow the team to evolve as the project progresses; also that its flexibility allows for risk to be allocated as appropriate and moved as the project develops. However, because the PPC 2000 is such a departure from the more traditional contracts, it has proved something of a struggle for those who use it for the first time. The contract can be difficult to interpret, as key elements such as risk sharing and key performance indicators are left as headings with the detail to be completed later. To guide new users, in June 2003 a Guide to ACA Project Partnering Contracts PPC 2000 and SPC 2000 was launched.

The key features of PPC 2000 include:

- A team approach with duties of fairness, team work and shared financial motivation.

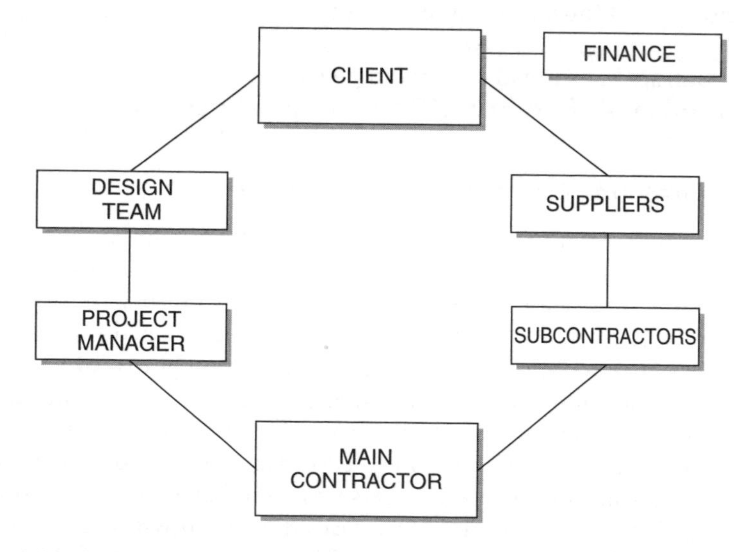

Figure 6.10 Costs and recommendations/multi-party PPC contract

- Stated partnering objectives – including, innovation, improved efficiency through the use of key performance indicators (KPIs) and completion of the project within an agreed time and to an agreed quality.
- A price framework which sets out the contractor's profit, central office overheads and site overheads as well as an agreed maximum price.
- A procedure for dispute resolution hierarchy.
- Commitment to the most advantageous approach to the analysis and the management of risk.
- The ability to take out latent defects and/or project insurance (see Chapter 1).

Some aspects may cause difficulties; for example, Section 3 of the contract gives the client representative the ability to inspect the financial records of any member of the team at any time subject to reasonable notice and access to members' computer networks and data by each member.

The SPC 2000 for specialist contractors

A further development is the publication of a subcontract to complement PPC2000 by the ACA. The specialist contract is intended to provide a standard document so that parties entering into PPC 2000 can have back-to-back arrangements with their subcontractors or specialists, to use the contract terminology. The SPC 2000 has the same basic structure as PPC 2000, but includes a specialist agreement to which the specialist terms are appended. The specialist contract endeavours to ensure that the constructor and the specialist work together more effectively than is perceived to be the case under the traditional forms of contract.

The NEC3 Partnering Agreement

To compare the PPC 2000 with the NEC3 Partnering Agreement is to compare apples with oranges. Whereas the PPC 2000 is a free-standing multi-party contract, governing all the parties' mutual rights and obligations in respect of a particular project, rather than being, as in the case of the NEC3 Partnering Agreement, an option bolted on to a series of bi-party contracts which must each be based upon the NEC3 form. When using NEC3 Option X12 each member of the partnering team has its own contract with the client (see Figure 6.11). The NEC3 Partnering Agreement, which by contrast to PPC 2000 is extremely short, acts as a framework for more detailed provisions which must be articulated by the parties themselves in the Schedule of Partners, or in the document called the Partnering Information. It is up to the parties to identify the objectives. Further provisions in the Partnering Agreement set out obligations which are an essential condition if those objectives are to be met (e.g. attendance at partners' and core group meetings, arrangements for joint design development, risk management and liability assessments, value engineering and value management).

Partnering using NEC3 Option X12

Prime contracting

Introduced in the 1990s, prime contracting is a long-term contracting relationship based upon partnering principles and is currently being used by several large public sector agencies as well as by some private sector clients. A prime contractor is defined as an entity that has the complete responsibility for the delivery and, in some cases, the operation of a built asset and may be either a contractor, in the generally excepted meaning of the term, or a firm of consultants. The prime contractor needs to be an organisation which has the ability to bring together all of the parties in the supply chain necessary to meet the client's requirements. There is nothing to prevent a designer, facilities manager, financier or other organisation from acting as a prime contractor. However, by their nature prime contracting arrangements tend to require the prime contractor not only to have access to an integrated supply chain with substantial resources and skills such as project management. To date most prime contractors are in fact large firms of contractors, despite the concerted efforts of many agencies to emphasise the point that this role is not restricted to traditional perceptions of contracting. One of the chief advantages for public sector clients with a vast portfolio of built assets is that prime contracting offers one point of contact/responsibility,

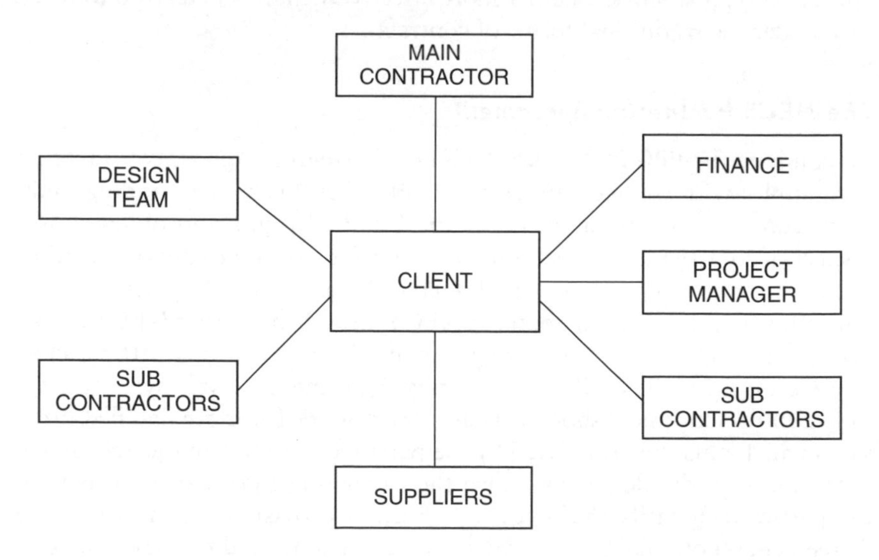

Figure 6.11 NEC3 Option X12

instead of a client having to engage separately with a range of different specialists.

For the client the major attraction is that the traditional dysfunctional system is replaced by single point responsibility. The prime contractor has the responsibility for:

- Total delivery of the project in line with through-life predictions, which can be up to thirty-five years from the time of delivery.
- Subcontractor/supplier selection – note there are exceptions to this rule (i.e. where a client may, because of its influence or market position, be able to procure some items more cheaply than the existing supply chain).
- Procurement management.
- Planning, programming and cost control.
- Design coordination and overall system engineering and testing.

One approach is to develop elements of the supply chain that constitute an integrated team which will work together on one particular part of the works, for example:

- groundworks
- lift installation
- roofing and so on.

These tiers are built outside of particular projects and there could be two or three supply chains capable of delivering an outcome for each tier. A typical tier for, say, mechanical and electrical services could include the design engineers, the contractor and the principal component manufacturers. Crucially, to achieve success in this approach, clusters must have the confidence to proceed in the design and production of their element in the knowledge that clashes in design or product development with other clusters are being managed and avoided by the prime contractor. Without this assurance this approach offers little more than the traditional supply chain management techniques where abortive and uncoordinated work is unfortunately the norm. It should be noted that the legal structure of such clusters has yet to be formalised.

The prime contractor's responsibilities might include the following:

- overall planning, programming and progressing of the work;
- overall management of the work, including risk management;
- design coordination, configuration control and overall system engineering and testing;
- the pricing, placing and administration of suitable subcontractors;
- systems integration and delivering the overall requirements.

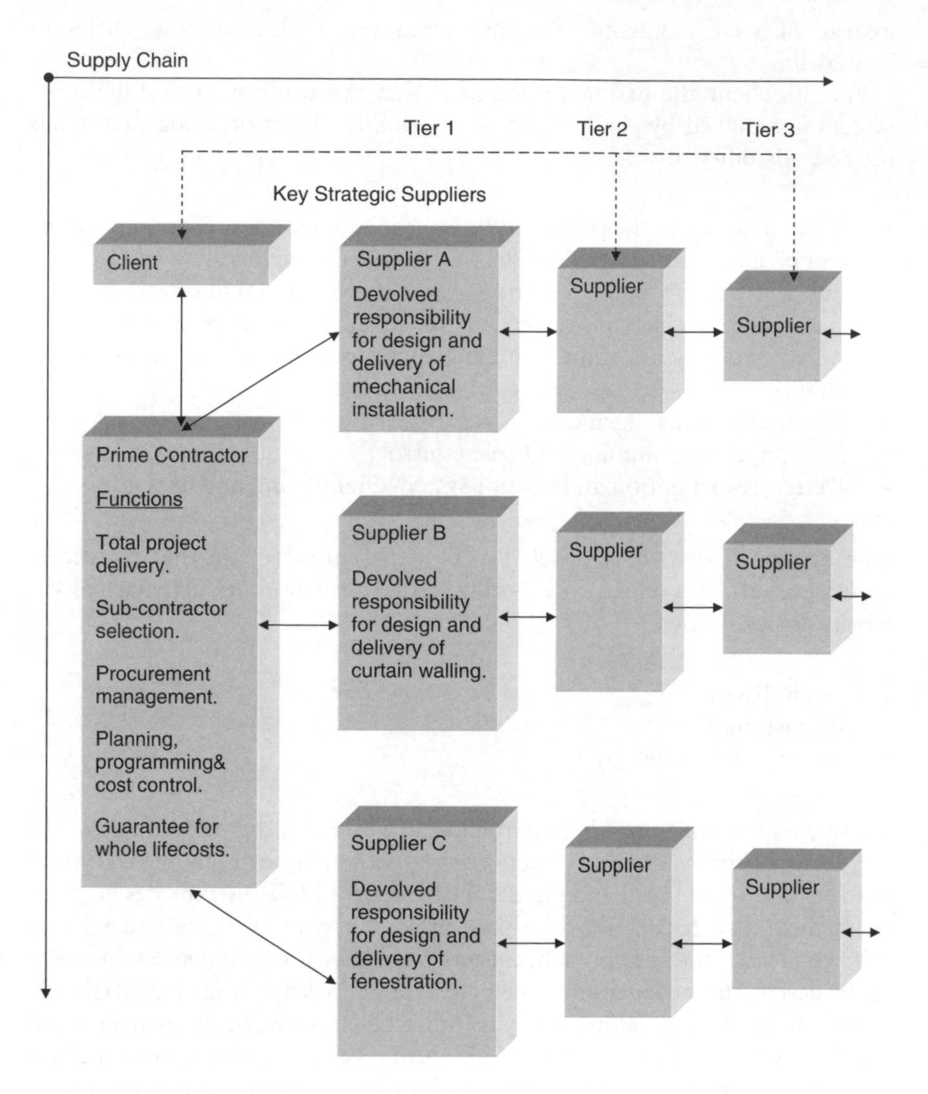

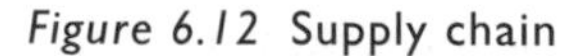
Step in agreement for client in areas of commercial benefit; e.g.- a super market chain's influence in the purchase of refrigeration equipment and plant.

Figure 6.12 Supply chain

The role of the quantity surveyor in prime contracting to date has tended to be as works adviser to the client, a role which includes examination of the information produced by the prime contractor, and one, it must be said, that is resented by prime contractors and as cost consultant to a prime contractor without the necessary in-house disciplines. However, there is nothing to prevent a quantity surveying consultancy from taking the prime contractor role, provided that the particular requirements of the client match the outputs, and that added value can be demonstrated.

Public/private partnerships (PPPs)

Background and definition

In its widest sense a public/private partnership (PPP) may be defined as ' a long term relationship between the public and private sectors that has the purpose of producing public services or infrastructure' (Zitron, 2004). One of the many PPP models (see Figure 6.13) is the Private Finance Initiative (PFI), a term used to describe the procurement processes by which public sector clients contract for capital-intensive services from the private sector. Private sector involvement in the delivery of public services in the UK has developed into a very emotive topic, with an unfortunate tendency to generate more heat than light. For many, the confusion and misconceptions surrounding PPP/PFI begin with the definition of these two terms. As discussed later in this chapter, the term PFI was launched in the early 1990s and then several years later the term PPP emerged and appeared to subsume the PFI. Public/private partnerships bring public and private sectors together in long-term contracts. PPPs encompass voluntary agreements and understandings, service-level agreements, outsourcing and the PFI.

A PPP project therefore usually involves the delivery of a traditional public sector service and can encompass a wide range of options, one of which is the PFI. In turn, the PFI is one of several similar approaches in

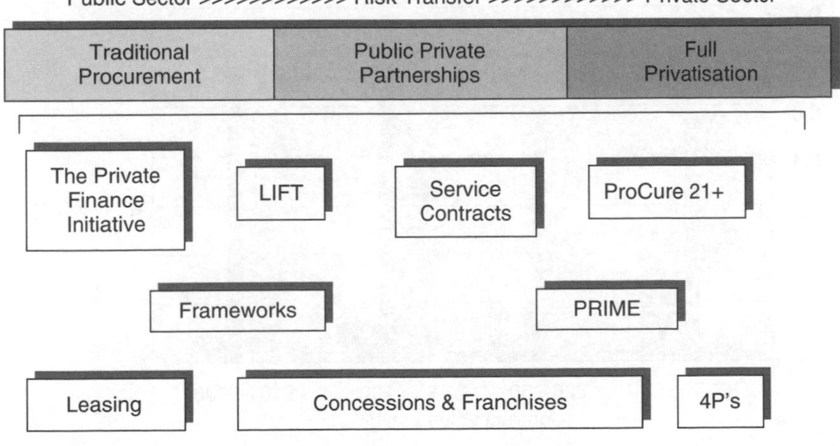

Figure 6.13 PPP procurement models

a 'family' of procurement that includes ProCure21 and LIFT (see Figure 6.13). It is claimed that one of the key objectives of the PFI is to bring private sector management expertise and the disciplines associated with private ownership and finance into the provision of public services. However, if the PFI is to deliver value for money to the public sector, the higher costs of private sector finance and the level of returns demanded by the private sector investors must be outweighed by lower whole-life costs and increased risk transfer.

The development of PPPs

In April 2004, in its Green Paper *On Public–Private Partnerships and Community Law on Public Contracts and Concessions,* the European Commission used the term 'phenomenon' to describe the spread of public/private partnerships (PPPs) across Europe. As will be discussed later in the chapter PPPs, and in particular the PFI, are now a global procurement model in which the UK is a world leader in terms of experience and know-how.

The origins of the UK Private Finance Initiative lie in the introduction, in 1981, of the Ryrie Rules, after Sir William Ryrie, a former Second Permanent Secretary of the Treasury. The Ryrie Rules were partially phased out in 1989 and finally abandoned in 1992 with the launch of the PFI.

The Private Finance Initiative is the name given to the policies announced by the Chancellor of the Exchequer, Norman Lamont, in the Autumn Statement of 1992. The Autumn Statements of 1993 and 1994 by Chancellor Kenneth Clarke were used to reshape the design and nature of the initiative. The intention was to bring the private sector into the provision of services and infrastructure, which had formerly been regarded as primarily a public

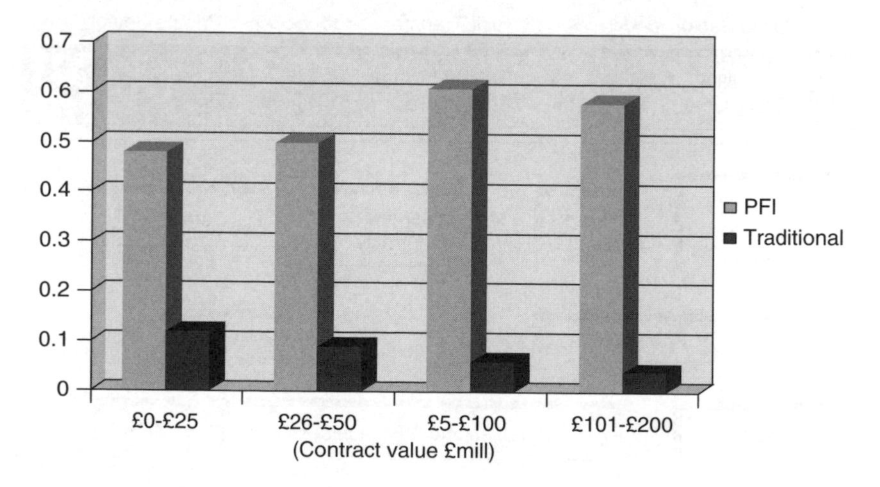

Figure 6.14 PFI Bid costs

sector concern. For many political spectators PFI was a natural progression for the Thatcher government that had so vigorously pursued a policy of privatisation during the 1980s. Not surprisingly therefore the PFI has been seen by some as a means of back-door privatisation of public services, and trade unions, in particular UNISON, have voiced their concerns over the adoption of the PFI. However, as far as government is concerned there is a clear distinction between the sale of existing public assets, which they see as privatisation, and the PFI, which they do not.

It was against this backdrop therefore that in 1992 the PFI was launched and almost immediately hit the rocks. The trouble came from two sides; first, the way in which civil servants had traditionally procured construction works and services left them without the experience, flexibility or negotiation skills to 'do deals', a factor that was to prove such an important ingredient for advancement of the PFI. In addition, there was still a large divide and inherent suspicion between the public and private sectors and very little guidance from government as to how this divide could be crossed.

The second major problem in trying to get the PFI off the ground related to the way in which a whole range of projects in the early days of the initiative were earmarked by overzealous civil servants as potential PFI projects, when they were quite obviously not. The outcome of this was that consortia could spend many months or even years locked into discussions over schemes with little chance of success, because the package under negotiation failed to produce sufficient guaranteed income to pay off the consortia's debt due to onerous contract conditions and inequitable risk transfer stipulations by the public sector. This practice earned PFI the reputation of incurring huge procurement costs for consortia and contractors before it became apparent that the business case for the project would not hold water. The procurement costs were non-recoverable by the parties concerned, and before long the PFI earned the reputation of being procurement of the last resort, at least

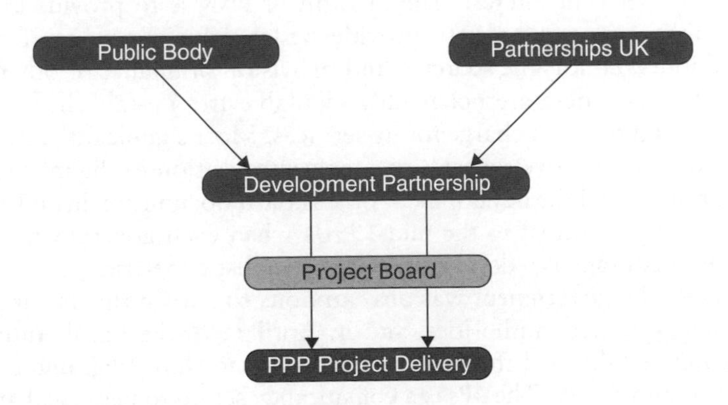

Figure 6.15 Development Partnership Agreement

by the private sector. Figure 6.14 illustrates the cost differential between traditional tendering and PFI procurement.

The 1997 Labour government was elected to power on a pledge to put partnership at the heart of modernising public services. Within a week of winning the election in May 1997 a Labour government appointed Malcolm Bates to conduct a wide-ranging review of the PFI. The first Bates Report made twenty-nine recommendations to which the dissolving of the Private Finance Panel in favour of a Private Finance Taskforce was key. Following the publication of the second Bates Report in 1999 and its recommendation that deal-making skills could be strengthened and that all public sector staff engaged in PFI projects should undergo annual training, Pricewaterhouse-Coopers was tasked with producing a PFI Competence Framework.

In 1999 Sir Peter Gershon was invited to review civil procurement in central government. The subsequent report highlighted a number of weaknesses in government procurement systems as follows:

- organisation
- process
- people and skills
- measurement
- contribution of the central government.

Gershon's aim was to modernise procurement throughout government, provide a greater sense of direction in procurement and promote best practice in the public sector. Gershon's proposals for dealing with these deficiencies led to the creation of a central organisation entitled the Office of Government Commerce (OGC).

In June 2000 Partnerships UK plc (PUK) was established (Figure 6.15) to replace the projects wing of the Treasury Taskforce as a joint venture between the public and private sectors with the private sector holding the majority 51 per cent interest. The mission of PUK is to provide expertise to the public sector in order to provide better value for money for PPPs. Included in its remit is the sourcing and provision of finance or other forms of capital where these are not readily available from established financial markets, and it makes a charge for its services. Most significantly, PUK may be seen to mark a move towards greater centralisation in the management of PPP projects and the development of standard documents, including contracts, in direct contrast to the mid-1990s when each government department was encouraged to develop its own specialist expertise.

However, the government was also anxious to spread the use of private investment into local authorities, and in April 1996 the Local Authorities Associations established the Public Private Partnership Programme or 4Ps in England and Wales. The 4Ps is a consultancy set up to help local authorities develop and deliver PFI schemes and other forms of public/private partnership. The local authority services covered by the 4Ps are, for example, housing, transport, waste, sport and leisure, education, etc.

During the second and third terms of the Labour government in the UK a number of specialist PPP procurement routes have been devised in order to meet the needs of particular public sector agencies.

NHS Local Improvement Finance Trust (LIFT)

Similar to LEPs, Local Improvement Finance Trusts (LIFTS) involve Partnerships UK plc (PUK) and the Department of Health forming a joint venture, Partnerships for Health, to encourage investment in primary care and community-based facilities and services. LIFT has been developed to meet a very specific need in the provision of primary and social healthcare facilities in inner city areas, that is to say GP surgeries, by means of a long-term partnering agreement. In order to participate in the programme, projects must be within areas designated as LIFT by the Department of Health. Although LIFT is at present confined to the health sector, other sectors are looking closely at the model for possible adaptation to other public service provision.

LIFT is based upon an incremental strategic partnership and is fundamentally about engaging a partner to deliver a stream of accommodation and related services through a supply chain, established following a competitive EU-compliant procurement exercise. Similar to the approach adopted by framework agreements, there should be no need to go through a procurement process again for a bidder to undertake these additional projects. Therefore, just as in the case of ProCure 21 (see below), there should be considerable savings in terms of cost and time over the duration of the partnership arrangement.

Frameworks

Framework agreements are being increasingly used to procure goods and services in both the private and public sectors. Frameworks have been used for some years on supplies contracts; however, in respect of works and services contracts, the key problem, particularly in the public sector, has been a lack of understanding as to how to use frameworks while still complying with legislation, particularly the EU Directives and the need to include an 'economic test' as part of the process for selection and appointment to the framework. In the private sector BAA was the first big player to make use of framework agreements and covered everything from quantity surveyors to architects and small works contractors. The EU Public Procurement Directives define a framework as 'An agreement between one or more contracting authorities and one or more economic operators, the purpose of which is to establish the terms governing contracts to be awarded during a given period, in particular with regard to price and, where appropriate, the quality envisaged.'

ProCure 21/ProCure21+

NHS ProCure 21 has been constructed by NHS Estates around four strands to promote better capital procurement by:

- Establishing a partnering programme for the NHS by developing long-term framework agreements with the private sector that will deliver better value for money and a better service for patients.
- Enabling the NHS to be recognised as a 'Best Client'.
- Promoting high-quality design.
- Ensuring that performance is monitored and improved through benchmarking and performance management.

In common with most large public sector providers the NHS has suffered from the usual problems of schemes being delivered late, over budget and with varied levels of quality combined with little consideration for whole-life costs. One of the main challenges to NHS capital procurement is the fragmentation of the NHS client base for specific healthcare schemes, since it comprises several hundreds of health trusts which all have responsibility for the delivery of schemes and each having differing levels of expertise and experience in capital procurement. The solution to these problems was to develop an approach to procurement known as NHS ProCure 21 as a radical departure from traditional NHS procurement methods and its cornerstone of the massive capital investment programme in the NHS in the period up to 2010. The principle underpinning the ProCure 21 programme is that of partnering with the private sector construction industry. From October 2010 ProCure21+ replaced the previous procurement model.

The PFI

The primary focus for PFI to date has been on services sold to the public sector. There are three types or PFI transactions currently in operation. The Private Finance Initiative is the widest used, most controversial and best-known form of PPP, currently accounting for approximately 80 per cent of all expenditure on PPPs in the UK construction sector. PFI deals have been used in some of the most complex and expensive PPP projects to date, such as the 872-bed New Royal Infirmary, Edinburgh (NRIE), and fall into three categories:

1. Classic PFI.
2. Financially free-standing projects.
3. Joint ventures.

1. Classic PFI

Typically the private sector finances, builds and then operates over a thirty- to sixty-year period a traditional public sector asset and in return receives a unitary payment based upon performance and availability. In some cases (e.g. prisons), the private sector also provides staffing. Using one of the most popular Private Finance Initiative models, namely Design, Build, Finance and Operate (DBFO) (see Figure 6.16), Consort Healthcare is a private sector consortium comprising service group BICC, the Royal Bank of Scotland and Morrison Construction which designed and built the NRIE between 1998 and 2002, including arranging and providing the debt finance. Since its opening in 2002 Consort Healthcare maintains non-clinical hospital services such as car parking, catering, cleaning, planned maintenance, etc. The public sector client, Lothian NHS Trust, retains the responsibility for the clinicians and clinical services, including all medical staff. In return for providing and running the hospital building and all the ancillary services, Consort Healthcare receives a predetermined performance-based unitary payment for the duration of the PFI contract (thirty years plus), providing, of course, that output and performance targets and standards are maintained, and the NHS Trust continues to enjoy a state-of-the-art hospital, including any commitments by Consort to refresh and update the technology and equipment during the contract period. The total value of the NRIE contract is £250 million over thirty-two years which includes not only the capital cost (£65 million) and finance costs but also the cost of the unitary charge. At the end of the contract period the hospital will be handed back, at no cost, to the NHS Trust in a good state of repair.

Classic PFI players

Special Purpose Vehicle/Company

A Special Purpose Vehicle is a consortium of interested parties brought together in order to bid for a PFI project. If successful, the consortium will be registered as a Special Purpose Company, usually at the time of financial close. The Special Purpose Vehicle/Company, or Shell Company as it is sometimes known, is a unique organisation constituted purely for a single PFI project. The company has a contractual link with the public sector sponsor or provider, the provider of finance, as well as the design, build and operating sectors of the project. In addition, the funder also usually has its own agreement with the sponsor which usually contains a step-in clause. This agreement is a safety net in the event that the Special Purpose Company ceases trading or persistently fails to deliver services to the required contract standards.

Public sector sponsor

The public sector client experience has shown that one of the key roles within the PFI procurement process is the client team's project manager. Optimum progress is made when the project is managed by a person who has the time and authority to take decisions and negotiate with bidders, instead of having to keep referring back. The public sector client also usually relies heavily on input from consultants in the fields of; procurement, including EU procurement law (see Chapter 5), project planning, production of an output specification, evaluation of bids, drawing up contract documents and business cases, etc.

The funder

One of the defining characteristics of PFI procurement is that the project is financed from private sector sources instead of central public funds. In the final stages of the PFI procurement the whole deal is put under the

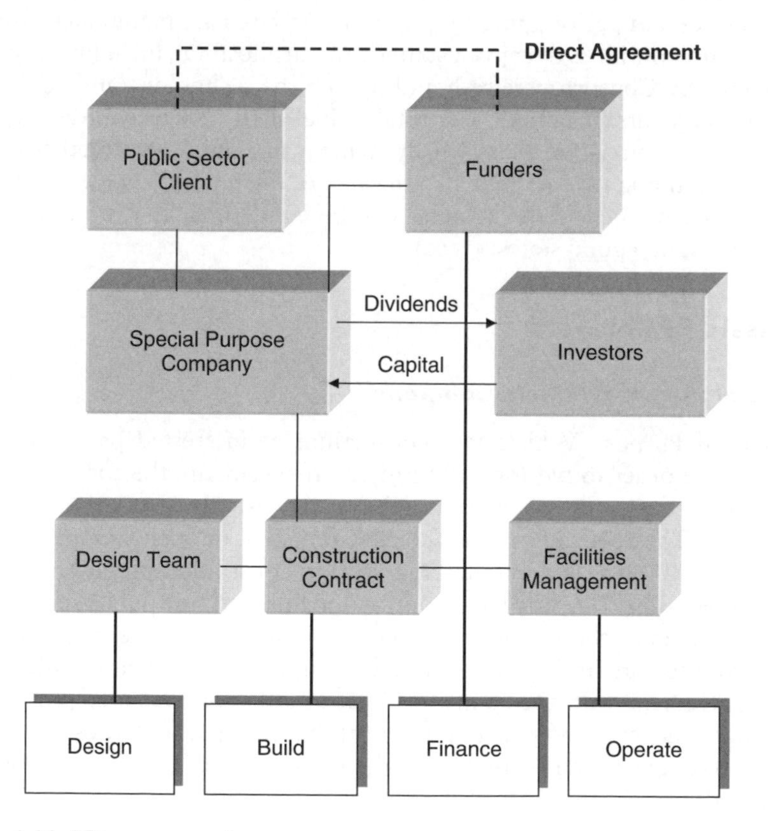

Figure 6.16 PFI contractual arrangements

microscope in a process known as due diligence, which is often carried out by the funders. These checks often pose serious questions about risk and other aspects of the contract that purchasers and providers think they have already resolved. This final step can be a trial of nerves as this process can take several weeks or even months, and can and frequently does involve previously agreed points being renegotiated to the satisfaction of the funders.

The design and construction

The design and construction part of the process is usually the most straightforward and easily understood part of the procedure, with the majority of design teams and contractors leaving the project once the construction phase is completed and ready to start operating. One of the major criticisms of PFI projects has been their lack of architectural merit and design innovation. Some of the causes given are that the design period is too short, and that there is too much commercial pressure and insufficient contact with the user/client.

The operator

The operator is the 'O' in DBFO and in the case of, say, a school, this organisation will have complete devolved responsibility for the day-to-day operation of the facility, which may include such diverse services as the provision and maintenance of information technology provision, lunches and playing field maintenance. In the case of ITCs there is often a contractual obligation to ensure that the hardware and software are kept up to date with the latest versions of programmes and technologies. The operator is clearly the major player in ensuring project success and receives payment based on the quality and reliability of the services.

2. Financially free-standing projects

The second PFI model is one where the private sector supplier designs, builds, finances and then operates an asset, recovering costs entirely through direct charges on the private users of the asset (e.g. tolling), rather than payments from the public sector. Public sector involvement is limited to enabling the project to go ahead through assistance with planning, licensing and other statutory procedures. There is no government contribution or acceptance of risk beyond this point and any government customer for a specific service is charged at the full commercial rate. Examples of this kind of project include the second Severn Bridge, the Dartford River Crossing and the Royal Armouries.

3. Joint ventures

Joint ventures are projects to which both the public and private sectors contribute, but where the private sector has overall control. In many cases the public sector contribution is made to secure wider social benefits, such as road decongestion resulting from an estuarial crossing. In other cases government may benefit through obtaining services not available within the timescale required. The project as a whole must make economic sense and competing uses of the resources considered. The main requirements for joint venture projects are:

- Private sector partners in a joint venture should be chosen through competition.
- Control of the joint venture should rest with the private sector.
- The government's contribution should be clearly defined and limited. After taking this into account, costs will need to be recouped from users or customers.
- The allocation of risk and reward will need to be clearly defined and agreed in advance, with private sector returns genuinely subject to risk.

The government's contribution can take a number of forms, such as concessionary loans, equity transfer of existing assets, ancillary or associated works, or some combination of these. If there is a government equity stake, it will not be a controlling one. The government may also contribute in terms of initial planning regulations or straight grants or subsidies.

Why PFI?

The uniqueness of PPPs lies in the partnership of two sectors (public and private) which have over the past sixty years or so, in the UK at least, followed very different paths, with very different objectives. In broad terms the benefit for the private sector includes the predictability of guaranteed long-term income streams and for the public sector, cost and time certainty in the delivery of a new or refurbished built asset that enables it to deliver a public policy outcome. In addition, given the unenviable track record of the UK construction industry, the public sector client does need to start paying for the facility or service until it is ready for use.

The difficulties with the traditional fragmented approach to public procurement have been threefold:

1. Projects can only proceed once the public funding is in place and this can be problematic. Agencies have to bid annually, recently changed

to three-yearly, for funds from the Treasury and inevitably many projects fail to secure funding and do not go ahead. If funding is secured, design and procurement is usually on the basis of cheapest bottom-line price rather than value for money, with little or no consideration given to long-term running, maintenance or decommissioning costs.
2. Once funding is approved the project delivery is often unreliable both in terms of cost and time certainty, as previously discussed in Chapter 1.
3. The maintenance of built assets is also dependent upon central government funding, which like the funding of capital projects is unpredictable. Often funds for capital building programmes have to be diverted to carry out essential maintenance or repair work.

In addition, traditional procurement models leave the public sector client vulnerable to high levels of risk which, it has been proved, it is ill-equipped to manage. PFI procurement results in a large proportion of risk being transferred to the contractor or private sector.

Compared with the traditional and often fragmented approaches to construction procurement PPPs, depending upon the model used, offer the advantages of synergies between traditionally diverse processes in the delivery and operation of built assets; for example:

- Synergy between the design and construction. This is not a new concept and buildability may also be achieved through other forms of procurement, such as design and build. Most PFI projects are able to deliver this well, with designers working alongside the contractor.
- Synergy between the construction phase and the operational phase. This is mainly to do with the suitability and reliability of the construction, taking into account whole-life costs over the expected life of the project.

Not unnaturally, there is growing evidence that companies that can combine, design, construction and hard facilities management in-house are increasingly successful in the PPP market (for example, the UK arm of the French giant Bouygues).

The current state of PPP/PFI

The near collapse of the world's financial markets in 2009 had a major impact upon the UK PFI market as sources of private funding dried up. Consequently deals slumped in 2009, making it the worst year since 1995 for the public/private partnerships industry. In 2009 only thirty-five PPP deals in total were signed across all sectors amounting to £4.24 billion; this compared to £6.5 billion in 2008 and £7.3 billion in 2007. Elsewhere in Europe the PPP market developed steadily, with around €5.0 billion worth of deals being signed; however, with unpredictable political and economic times ahead in the UK it is uncertain what the level of activity will be during the coming years.

In Scotland the Scottish National Party announced upon their election in 2007 that the PFI was no longer to be an option as a public sector procurement route in Scotland; instead it was to be replaced by the Scottish Future's Trust in 2008 but is not due to initiate procurement programmes until 2010/2011. With current PFI projects coming to a close, it is believed by some that this will result in a loss of PFI expertise north of the border.

In June 2010 a Conservative/Liberal coalition government was elected to Westminster. Within weeks, in an attempt to cut costs, the Building Schools for the Future programme with a budget of £9 billion over three years was axed. The future of other PPP programmes is, at the time of writing, uncertain.

So what is the current state of health of the PFI and why is it used? When it was first launched in 1992 the principle rationale was to:

- provide value for money and efficiency savings, as well as
- to transfer risk from the public to the private sector.

These motives still remain largely the driving force as the procurement policy matures.

The PFI procurement process – 'Getting a good deal'

Figure 6.17 illustrates the recommended Treasury Taskforce procurement path for PFI projects. The PriceWaterhouseCoopers report 'PFI Competence Framework' suggested that these stages fall into three broad phases:

Phase 1 Feasibility – Stages 1–4.
Phase 2 Procurement – Stages 5–13.
Phase 3 Contract management Stage 14.

During the past five years or so there have been various attempts to modify the procurement process as it has been criticised for lacking flexibility and being too long. For example, in the PFI project for the redevelopment of West Middlesex Hospital which opened in 2004 a round of bidding was omitted in order to speed up the process. Subsequently the National Audit Office concluded that the Trust ran an effective bidding competition but that it should be noted that if this strategy was to be used in future PFI deals then the following safeguards need to be put in place to maintain competitive tension when using this approach. It is recommended that the public sector client should:

- obtain greater bid details at an early stage
- keep the main aspects of the deal constant in the closing stages
- be prepared to walk away from the preferred bidder
- make it clear to bidders that this process is to be applied
- ensure that there are no major open issues for negotiation.

The stages that have proven to be of crucial importance in determining the case for the use of PFI are:

- *Stage 3* – An identification of key risks. Risk transfer is one of the key tests for a good PFI deal as value for money can be demonstrated to increase each time a risk is transferred. There are two aspects to risk transfer:

    1. Between the public and private sector.
    2. Between the members of the PFI consortium.

In most PFI projects the risks that are earmarked for transfer to the private sector are by now fairly standardised and well understood; however, major difficulties can arise in deciding who within the consortium carries the various burdens of risk. In the case of risk transfer between public and private sectors the main drivers are transparency and the need to demonstrate value for money; in the case of risk transfer within the consortium the commercial interests of the various players – that is to say; financial institutions, contractors, operators – dominate the discussion. The principle governing risk transfer is that the risk should be allocated to whoever from the public or private sector is able to manage it at least cost; that is to say, identified risks should be either retained, transferred or shared. The valuation of risk transfer, however problematic, often tips the scales on PFI deals as the public sector comparator alone often shows that value for money has not been demonstrated.

- *Stage 9* – Refining the proposal.
- *Stage 12* – Selection of the preferred bidder.

Public/private partnership projects (4Ps)

This form of PPP was introduced in 1996 with the intention of encouraging local authorities and councils to consider PPPs for the delivery of some services. 4Ps projects are long-term contract of thirty years plus and are really a form of PFI, in which central government goes through various departments.

PFI Procurement Guide

STAGE 1	ESTABLISH BUSINESS CASE – It is vitally important that the PFI project is used to address pressing business needs. Consider key risks.
STAGE 2	APPRAISE THE OPTIONS – Identify and assess realistic alternative ways of achieving the business needs.
STAGE 3	OUTLINE BUSINESS CASE – Establish the project is affordable and 'PF lable.' A reference project or Public Sector Comparator should be prepared to demonstrate value for money including a quantification of key risks. Market soundings may be appropriate at this stage – see Chapter 6.The outline Service Specification should be prepared.
STAGE 4	DEVELOPING THE TEAM – Form procurement team with appropriate professional and negotiating skills.
STAGE 5	DECIDING TACTICS – The nature and composition of the tender list and selection process.
STAGE 6	PUBLISH OJEU – Contract notice published in OJEU – See Chapter 6 and Appendices.
STAGE 7	PREQUALIFICATION OF BIDDERS– Bidders need to demonstrate the ability to manage risk and deliver service.
STAGE 8	SELECTION OF BIDDERS – Short-listing. Method statements and technical details may be legitimately being sort.
STAGE 9	REFINE THE PROPOSAL – Revise its original appraisal (Stage 3) and refine the output specification, business case and public sector comparator.
STAGE 10	INVITATION TO NEGOTIATE – Could include draft contracts. Quite lengthy – 3 to 4 months. Opportunity for short listed bidders to absorb contract criteria and respond with a formal bid.
STAGE 11	RECEIPT AND EVALUATION OF BIDS – Assessment of different proposals for service delivery.
STAGE 12	SELLECTION OF PREFERRED BIDDER – Selection of preferred bidder with bid being tested against key criteria.
STAGE 13	CONTRACT AWARD AND FINANCIAL CLOSE – Sign contract and place contract award notice in OJEU. See Chapter 6.
STAGE 14	CONTRACT MANAGEMENT – Operational and management relationship between public and private sectors.

Figure 6.17 PFI procurement guide

Source: HM Treasury.

The role of the quantity surveyor in PPPs/PFIs

Quantity surveyors are just one of a range of professional advisers involved in PPP/PFI projects. As previously noted, the Royal Institution of Chartered Surveyors is in no doubt as to the importance of PPPs/PFIs in the future of the profession and many quantity surveying practices are involved in PPP/PFI deals in a variety of roles for both the public and private sectors. Figure 6.18 illustrates the public/private skills balance as identified by the RICS Project Management Faculty. In the private sector when working for the operator the quantity surveyor's role may involve the various PPP skills listed in Figure 6.18.

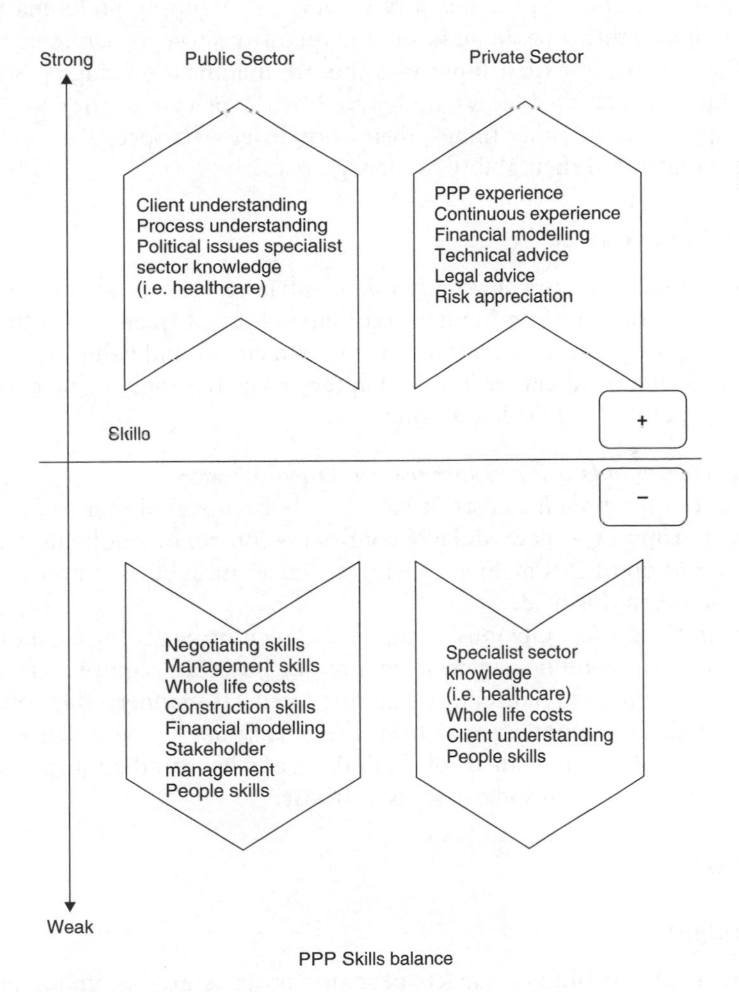

Figure 6.18 PPP skills balance

Source: RICS Project Management Progressional Group.

The private sector – special purpose companies

Advice on procurement

For many private sector consortia the approach to submitting a bid for a PFI project is unknown territory. Added to this is the fact that by their very nature PFI projects tend to be highly complex, requiring decisions to be taken during the development of the bid at Stage 8 that not only involve capital costs, but also long-term costs. Increasingly, as explained earlier in this chapter, the impact of EU procurement directives must be considered. Some contracting authorities use the OJEU to 'test the water' for a proposed PPP project at Stage 6; the quantity surveyor can supply preliminary cost information at this time. In addition, the quantity surveyor with experience in PPP can provide expert input into the pre-qualification stage – Stage 7. The stage at which the bidders are selected by the private sector consortia is based upon, among other things, their knowledge of a specialised sector of public services and their ability to manage risk.

General cost advice

The traditional quantity surveying role is advising on capital costs, including the preparation of preliminary estimates, bills of quantities, obtaining specialist quotes, etc. In addition, value management and value engineering techniques described earlier in this chapter are increasingly being called on to produce cost-effective design solutions.

- *Reviewing bids prior to submission.* Due diligence.
- *Advice on whole-life costs.* It has already been stated that to many the key to running a successful PPP contract is control of whole-life costs. In recognition of this many surveying practices now have in-house advice available in this field.
- *Specialist advice.* Obviously, highly complex projects, for example, the construction and management of a major hospital, require a great deal of input from specialists (for example, medical planners able to advise on medical equipment, etc.) from the outset. Surveying practices committed to developing their role in PPP already have at their disposal such expertise, which in some cases is in-house.

Funders

Due diligence

The financial and funding aspects of major projects are becoming increasingly susceptible to both technical as well as commercial risks. Investors and funding institutions are looking more and more for independent scrutiny of all aspects of development from, design integrity to contractual robustness of the contract and beyond to the expenditure levels and progress against programmes. The skills of the quantity surveyor provide an excellent platform for the investigative and analytical processes necessary to satisfy these requirements.

The public sector purchaser

Procurement advice

For many in the public sector, this method of procurement is just as unfamiliar as the private sector. The surveyor can advise the contracting authority on how to satisfy the requirement of this method of procurement. It is widely agreed that the appointment of a project manager at an early stage is vital to PPP project success. In addition, pressure is being exerted to speed up the procurement process, a factor that makes the role of the project manager even more crucial.

- *The outline business case (OBC).* The preparation and development of the OBC in Stage 3 involves the preparation of a risk register, the identification and quantification of risk; all of which are services that can be supplied by the quantity surveyor.
- *Advice on facilities management.* Technical advice on this topic during the drawing up of the service specification during Stage 3 and beyond.

Common and joint services to SPCs/public sector purchasers

Joint public/private monitor certifier

This role is similar to the role played by a bureau de controle in France and involves monitoring the construction work to ensure that it complies with contract. In addition to the built asset the surveyor employed in this role can monitor facilities management operation. The concerns with this practice centre around the 'belt and braces' way in which the certification is being carried out and the fact that firms are signing off multi-million-pound schemes for a very small fee, and are effectively acting as unpaid insurance agents, with any claim being covered by professional indemnity insurance.

Services to consortium building contractors

The role recognised by many surveyors as their main involvement in PPP, it includes preliminary cost advice, preparation and pricing of bills of quantities, and supply chain management.

Conclusions

There is still a heated debate at all levels as to whether PPPs and in particular PFIs deliver value for money nearly twenty years after this procurement route was introduced in its current model. In 2009 a House of Lords Select Committee concluded that the PFI was just about delivering value in terms of the provision of new public facilities, but there are others, for example, Professor Allyson Pollock, who gave evidence to the House of Lords and who will never be convinced of the benefits that PFI has brought to many public services.

Bibliography

Audit Commission (2001). *Building for the Future, The Management of Procurement under the Private Finance Initiative*, London: HMSO.

Catalyst Trust (2001). *A Response to the IPPR Commission on PPPs*, London: Central Books.

CGR Technical Note 1 (2004). London: HM Treasury.

Ernst & Young (2003). *PFI Grows Up*, London: Ernst & Young.

European Commission (2003). *Guidelines for Successful Public Private Partnerships*, Brussels: European Commission.

European Commission (2004). *On Public Private Partnerships and Community Law on Public Contracts and Concessions*. Brussels: European Commission.

Flyvbjerg, B., Bruzelius, N. and Rothengatter, W. (2003). *Megaprojects And Risk – An Anatomy Of Ambition*, Cambridge: Cambridge University Press.

Gosling, T. (ed.) (2004). *3 Steps Forward, 2 Steps Back*, London: Institute for Public Policy Research.

HM Treasury (1998a). *Stability and Investment for the Long Term*, Economic and Fiscal Strategy Report – Cm 3978 (June), London: HMSO.

HM Treasury (1998b). *Policy Statement No. 2 – Public Sector Comparators and Value for Money*, London: HMSO.

HM Treasury (1999). *Technical Note No. 5 – How to Construct a Public Sector Comparator*, London: HMSO.

Institute for Public Policy Research (2001). *Management Paper, Building Better Partnerships*, London: IPPR.

Kelly, G. (2000). *The New Partnership Agenda*, London: The Institute of Public Policy Research.

Meeting the Investment Challenge (2003). London: Office of Government Commerce.

Ministry of Defence (1999) *Building Down Barriers*, London: HMSO.

Ministry of Finance (2000). *Public–Private Partnership – Pulling Together*, The Hague: PPP Knowledge Centre.

National Audit Office (1999). *Examining the Value for Money of Deals under the Private Finance Initiative*, London: HMSO.

National Audit Office (2001a). *Innovation in PFI Financing, The Treasury Building Project*, London: HMSO.

National Audit Office (2001b). *Managing the Relationship to Secure a Successful Partnership in PFI Projects*, London: HMSO.
National Audit Office (2001c). *Modernising Construction*. London: HMSO.
Office of the Deputy Prime Minister (2004). *Local Authority Private Finance Initiative: Proposals for New Projects*, London: HMSO.
Partnerships for Schools (2004). *Building Schools for the Future*, London: HMSO.
PricewaterhouseCooper (1999). *PFI Competence Framework – Version 1* (December), London: Private Finance Unit, HMSO.
PricewaterhouseCooper (2001). *Public Private Partnerships: A Clearer View* (October), London: Private Finance Unit, HMSO.
Robinson, P. *et al.* (2000). *The Private Finance Initiative – Saviour, Villain or Irrelevance,* London: The Institute of Public Policy Research.
Royal Institute of Chartered Surveyors (2005). *Beyond Partnering: Towards a New Approach in Project Management*, London: RICS.
Williams, B. (2001). *EU Facilities Economics*, London: BEB Ltd.
Zitron, J. (2004). *PFI and PPP: Client and Practitioner Perspectives. Proceedings of 21st Annual Conference of the Major Projects Association*, London: Major Project Association.

Journals

Buckley, C. (1996). Clarke and CBI unite to revive PFI. *The Times*, 22 October.
Government Opportunities (2001). *Challenges to Procurement: The IPPR and Byatt Reports,* Norman Rose, 24 July.
Robinson, P. (2001). PPP tips the balance. *Public Service Review* PFI/PPP 2001, Public Service Communication Agency Ltd, UK.
Thomas, R. (1996). Initiative fails the test of viability. *Guardian*, 22 October.
Wagstaff, M. (2005). The case against. *Building Magazine*, 14 January.
Waites, C. (2001). Are we really getting value for money? *Public Service Review* PFI/PPP 2001, Public Service Communication Agency Ltd, UK.

Websites

www.dfee.gov.uk
www.doh.gov.uk
www.hm-treasury.gov.uk
www.minfin.nl/pps
www.nao.gov.uk
www.mod.gov.uk
www.ppp.gov.ie
www.4Ps.co.uk

Chapter 7

Emerging practice

Traditional quantity surveying practice may be said to be closely linked to trends in procurement practice as well as market conditions. The 2007 RICS survey of contracts in use captured a smaller number of projects than previous surveys but was able to reveal the following trends:

- The majority of building contracts in this country continue to use 'traditional' procurement; that is to say a transparent paper-based system. The use of electronic tendering and procurement systems remains at low levels despite the encouragement given by professional bodies such as the RICS to adopt paperless systems.
- The industry continues to show its conservatism, with the vast majority of building projects using a standard form of contract and JCT contracts continue to be the preferred family of use. Despite the fact that the NEC suite of contracts has been widely available for a number of years and strongly recommended by Latham and Egan, uptake continues to be slow, although there are signs that the NEC is gaining in popularity.
- The dominant procurement strategy is now design and build, although that is not universally the case, with bills of quantities still remaining popular in Scotland as well as in the public sector, where accountability is important. As discussed in Chapter 2, the much anticipated RICS New Rules of Measurement should finally hit the streets in 2011. It remains to be seen how long the various editions of the Standard Method of Measurement remain in use. The move to design and build reflects building clients' desire to shift risk to the client. Over 50 per cent of contracts in the £10,000 to £50 million value bands were procured on a design-and-build basis.
- The industry does not seem to be impressed with partnering with little increase year on year of this approach and has little use of partnering arrangements with standard forms of contract.

As discussed in the opening chapter, as fee scales for traditional services decline, and clients demand a greater range of advice, quantity surveyors are looking for and acting in different specialist fields, some of which are listed below. Although of course it is not simply a question of adding a new service to the letter heading, the questions of competency and indemnity insurance cover must always be considered.

- development management
- project monitoring
- employer's agent
- CDM coordinator
- disaster management.

Development management

The RICS *Guidance Note on Development Management* defines the role as 'The management of the development process from the emergence of the initial development concept to the commencement of the tendering process for the construction of the works.'

The role of the development manager therefore includes giving advice on:

- development appraisals;
- planning application process;
- development finance;
- procurement.

There are several definitions of the term development manager, and Table 7.2 offers a comparison of the development management process as defined by:

- The RICS Guidance Note.
- The CIOB's Code of Practice for Project Management for Construction and Development.
- Construction Industry Council (CIC) Scope of Services (major projects).

Table 7.1 Specialist fields

Types of services	*Cost advice*	*Procurement and tendering*	*Contract administration*	*Commercial management*	*Whole-life costs*
Quantity surveying	✓	✓	✓	✓	✓
Project management			✓		
Development management	✓	✓			✓
Project monitoring			✓	✓	
Employer's agent		✓	✓		

Table 7.2 RICS Guide on Development Management (2009)

Development management process	*RIBA's outline plan of work*	*CIOB's Code of Practice for project management for construction and development*	*CIC's scope of services (major works)*
Phase 1 – Developers' initial concept	A. Appraisal	1. Inception	1. Preparation
Phase 2 – Site acquisition strategy	A. Appraisal	2. Feasibility (site selection and acquisition)	1. Preparation
Phase 3 – Outline appraisal	B. Strategic brief and C outline proposals	3. Feasibility	2. Concept
Phase 4 – Outline planning permission	C. Outline proposals	4. Strategy	2. Concept and part of 3 design development
Phase 5 – Full planning permission	D. Detailed proposals	5. Pre-construction	3. Design development

Development appraisals

It is important that development appraisals and valuations are determined and carried out in accordance with the RICS Valuation Standards referred to colloquially as the Red Book.

Project monitoring

Project monitoring is distinct from both project management and construction monitoring, and is defined in the RICS *Project monitoring guidance note* (2007a) as 'Protecting the Client's interests by identifying and advising on the risks associated with acquiring an interest in a development that is not under the Client's direct control for example a Private Finance Initiative project'. The recent boom years that saw financial institutions lending large sums of money to individuals who were ill-equipped with the necessary skills has led to an increasing number of high-risk, distressed projects.

Project monitoring may be carried out for a range of clients, including:

- A funding institution which acquires the scheme as an investment on completion.
- A tenant or purchaser who enters into a commitment to lease or purchase a property on completion.

- A bank where a loan matures at the end of the development period.
- Grant funders.
- Private finance initiative funders and end-users. Note: when used on PFI contracts the advice also included a commentary on the whole-life costs for the period of the concession agreement.

The role of a project monitor is principally on a design and build project where he or she monitors the performance of the developer and its team and is investigator and adviser to the client.

Types of project monitoring

Matters relating to:

- land and property acquisition
- statutory compliance
- competency of the developer
- financial appraisals
- legal agreements
- construction costs and programmes
- design and construction quality.

The key stages in project monitoring are said to be:

- initial audit role
- progress reporting
- practical completion.

Benefits to the client are:

- enhanced risk management
- enhanced financial management
- enhanced programme management
- enhanced quality management.

Employer's agent

The most common situation where an employer's agent is used is when the JCT (05) design-and-build contract is used. Design and build and its variants has over the past twenty years overtaken traditional lump sum contracts based on JCT(05) as the popular form of contract among clients with more that 40 per cent of contracts being let on this basis (RICS and Langdon 2009). The JCT (05) design-and-build contract envisages the employer's agent undertaking the employer's duties on behalf of the employer and is provided for in Article 3. It is important to remember that an employer's agent is employed to administer the conditions of contract but does not perform the same function as the architect, contract administrator or project manager. For the chartered surveyor therefore the exact position of the employer's agent can be confusing and in particular the duties, if any, that they owe to the contractor: 'and otherwise act for the employer under any of the conditions'.

The employer's agent has a duty to act in a manner that is independent, impartial and fair in situations where the EA is required to make decisions on issues in which the employer's and contractor's interests may not coincide. In order to try to clarify the position of the employer's agent, the RICS has prepared a schedule with a list of potential EA services, with a tick box against each service. The schedule is suitable for use with the RICS Standard and Short Forms of Consultant's Appointment. Acting on behalf of the client/employer in respect of administration of a 'design-and-build' contract incorporating issue of notices and certificates in respect of:

- Preparation of employer's requirements documentation in association with the client and other consultants.
- Instructions in respect of expenditure of provisional sums, interim payment certificates for valuations of works and materials on and off site.
- Instructions in respect of variations, changes, confirmation of information and consents, opening up works for inspection, instructing procedure to be adopted in respect of antiquities found on site, advising on conflicts within the contract documentation, value instructions including the effects of postponement of design and/or construction works, including any loss and/or expense and the like.
- Statements identifying the part or parts of the works taken into early possession by the employer.
- Non-completion certificates.
- Certificates of practical completion and accompanying schedules of defective works.
- Certificates of making good defects at the end of defects liability period or at completion of defects (whichever is the latter).
- Final payment certificate following agreement of final statement of account with contractor and certificate at the end of the defects liability period.
- Costs and expenses properly incurred by the employer should either party determine the contract.

As with general project management appointments, there is no commonly accepted standard role and service for the role of employer's agent. However, assuming a broad role both pre- and post-contract, the following could form the basis of an agreed role.

Pre-construction

Employer's agents can assist with all facets of pre-construction planning, from assessment of the brief and communicating this to team members, to identifying the project execution plan.

The scope of service for pre-construction also includes:

- appointment of the team to suit the procurement and programme constraints;
- production of a comprehensive master programme;
- development of a risk register;
- establishing a management framework;
- development of an information required schedule and design strategy;
- manage the signing-off of the master plan, master programme, project brief and delivery strategy;
- change control procedures;
- delivery of a procurement route strategy.

Construction

- Pre-qualification and selection of contractor(s).
- Preparation of the employer's requirements if a design-and-build route is selected.
- Management of the procurement process.
- Refinement of the construction methodology, employer's requirements and contractor's proposals.
- Administration of the terms of the contract.
- Monitoring the site performance of the contractor to ensure that key milestone dates are achieved.
- Management of the phased completion of the project.
- Management of the flow of information between the contractor and the design team.

Post-construction

- Manage the process of issuing of operation and maintenance (O&M) manuals, and ensure adequate training is given to any facility management.
- Manage the post-construction process to ensure that any post-PC issues are rectified.

A true employer's agent must not act unreasonably, dishonestly or capriciously in withholding approvals or certificates but has little or no discretion and must obey the employer's instructions. Sometimes the role of employer's agent is combined with the role of certifier, who has much wider discretion in performing their duties. A certifier must form and act on their opinion when performing the role. Combining the two roles can create potential conflict.

The true employer's agent is a creation of the JCT design-and-build contract where the contract envisages that the employer's agent undertakes the employer's duties on behalf of the employer. Article 3 gives the EA the full authority to receive and issue:

- applications
- consents
- instructions
- notices
- requests or statements
- otherwise act for the employer.

The employer's agent has no independent function, but is the personification of the employer; they must act as instructed by the employer and has no discretion.

The employer's agent must take care to act within the terms of the authority given to them by the employer.

EAs can be put in a difficult position in the case where an employer is acting unreasonably, whereas the EA is required to act fairly and honestly.

The certifier

An employer's agent may also act as a certifier which is different to and separate from the role of the EA.

The employer's agent has very little discretion in carrying out their duties. However, once the role extends to include issuing certificates or approvals and requires the exercise of discretion and professional expertise the position becomes more complicated.

The CDM coordinator

This role requires technical knowledge of many aspects of the industry, an understanding of the design and construction process and the ability to communicate effectively. The CDM regulations are primarily concerned with health and safety on the construction site. Table 7.3 outlines the key aspects of the CDM coordinator's service.

Disaster management

Following the tsunami disaster in the Indian Ocean in 2004 the RICS established the President's Major Disaster Commission (MDMC), and

Table 7.3 Key aspects of the CDM coordinators' service

Advice and assistance	Providing proactive advice and practical help to the client in response to client and project demands.
Advising on competence of client appointments when necessary	Providing specific advice, systems or support to the client on how to comply with Regulation 4 and Appendix 4 of the ACoP relating to health and safety resources and competence.
Coordination and cooperation	Ensuring that suitable arrangements are made and implemented for the coordination of health and safety measures during planning and preparation for the construction phase. This process involves an active contribution.
Management arrangements	Supporting the client in identifying and ensuring suitable arrangements for the project, how they will be delivered by the team to achieve project safety and other related client-project benefits and how they will be reviewed and maintained throughout the life of the project. Some clients have arrangements in place already, which may require less advice and assistance from the CDM coordinator.
Information management	Developing a strategy with the team for maintaining the flow of relevant health and safety-related information throughout the lifetime of the project to make sure that what is needed reaches the right people at the right times. This includes information required by designers, pre-construction information, whenever it is required. and information for health and safety file.
Design risk management	Promoting the suitability and compatibility of designs and actively seeking the cooperation of designers at all project phases when dealing with the risk consequences of construction and workplace design decisions.
The start of the construction phase when required	Providing support for the client and advising on the suitability of the principal contractor's construction phase plan. The client will be entitled to rely on the CDM coordinator's advice at this transitional phase, as is a focus on the main objectives of planning and preparation for project safety.
Construction liaison and involvement	Encouraging and developing links between permanent and temporary works designers and actively liaising with principal contractors to ensure safe design.

Source: Health and Safety Executive (HSE).

subsequently the RICS in collaboration with other professional bodies such as the RIBA and the ICE published *The Built Environment Professions in Risk Reduction and Response* in 2009.

The guide outlines the responsibilities and capabilities of engineers, architects and surveyors and their professional institutions in reducing risks and responding to disaster impacts. Disaster is defined in the guide as 'the impact of different physical, social, economic, political and complex hazards on vulnerable communities' and includes a wide range of events, from earthquakes, war and civil conflict to development projects such as large dams.

Chartered quantity surveyors are able to contribute a variety of skills, including:

- measurement
- whole-life cycle advice
- investment appraisals/procurement advice
- managing the whole construction process.

An example of how a quantity surveyor may contribute to disaster management is in the area of risk management and response.

Risk and vulnerability assessment

Risk and vulnerability assessment involves identifying the nature and magnitude of current and future risks from hazards to people, infrastructure and buildings, and particularly vital facilities such as hospitals and schools. Risk can be assessed using computer modelling of natural disasters employing satellite image-based mapping. This can be combined with consultation with communities concerning their vulnerability and ability to cope with a hazard, particularly when climate change may threaten precarious land rights.

Disaster risk reduction and mitigation

Preparing a strategy to reduce vulnerability against known risks is a complex and continuous exercise involving strengthening vulnerable structures, preventing building activity in high-risk areas, managing and maintaining assets, and ensuring the enforcement of building regulations. Community-based disaster preparedness (CBDP) is already in place in many parts of the world and is a reliable vehicle for disaster prevention at the grassroots level, particularly for vulnerable groups such as the young, the elderly and the infirm.

Bibliography

Harvey, J. (2004). *Urban Land Economics – Sixth Edition*. Palgrave Macmillian.

Hillebrandt, P.M. (2000). *Economic Theory and the Construction Industry*. Palgrave Macmillian.

RICS (2007a). *Project Monitoring Guidance Note*. RICS Books.

RICS (2007b). *Quantity Surveyor Services*. RICS Books.
RICS (2009). *Guide on Development Management, First Edition Guidance Note*. RICS Books.
RICS and, Langdon D. (2009). *Contracts in Use: A Survey of Building Contracts in Use 2007*. RICS Books.

Chapter 8

Sustainability, assessment and quantity surveying practice

Rohinton Emmanuel

Introduction

A key new trend in the practice of quantity surveying in our time is the need to account for sustainability in the built environment. This need varies from facilitating the achievement of certified levels of sustainability in the built environment to incorporating sustainability value into building procurement and best value. The past two decades have seen myriad sustainability assessment protocols and systems sprung up from practically all corners of the Earth, yet assessment largely remains an esoteric 'nice-to-have' addition rather than something integral to the development, design, construction, operation and end-of-life dismantling of built assets (Bouwer *et al.*, 2005; Cole, 2007; Forum for the Future, 2007). A critical and glaring gap is the near total absence of a system of sustainability value that is transparent and practical to the day-to-day practice of the profession. In this chapter I will present the state of the art in sustainability assessment in the built environment and make a case for three broad areas in which the core skills of quantity surveyors can play a central role in achieving sustainability in the contemporary built environment.

Sustainability in the built environment and the role of quantity surveyors

The design, construction, use, refurbishment and dismantling of built assets is a major consumer of resources and generates considerable waste. The construction sector accounts for approximately one-third to half of all commodity flows in countries that are members of the Organisation for Economic Cooperation and Development (OECD) (OECD, 2003). The construction sector is responsible for nearly 40 per cent of all solid waste generated in OECD countries (OECD, 2008) and on average, 25 to 40 per cent of national carbon emissions come from energy used to heat, cool, light and run buildings. Table 8.1 highlights the global resource consumption and pollution loading from the construction sector.

Among the building stock, housing plays a key role in carbon emission: in the UK; 27 per cent of national carbon emission comes from housing (of which nearly half is from space heating), while water heating and appliances/lighting/cooking account for 20 per cent and 25 per cent respectively (Monk *et al.*, 2010). While the environmental consequences of the built environment are significant, construction as a sector is too big to ignore in the wider attempt to make the total economy sustainable. At the EU-27 level, the construction industry represents nearly 10 per cent of GDP (€1,305 billion) and accounts for 7.3 per cent of the total workforce (13.2 million people) (European Commission, 2007a). Construction materials and building products account for one-third of EU economic activity (European Commission, 2007a). In the UK, construction represents nearly 9 per cent of the gross domestic product (approximately £114 billion of gross value-added in 2008), and employs 2.6 million people (IGT, 2010).

In the social sphere, buildings contribute positively to people's quality of life, providing the surroundings in which we work, rest and play, are educated and are healed. At the same time, many built environments are blighted by crime and poverty. Poorly designed and procured buildings have negative effects on the health, welfare and economic prospects of companies and communities.

Given the central role of the construction sector in both positively influencing quality of life with potential negativity towards social and environmental factors, what can we say about the role which quantity surveyors are playing or should play in firmly anchoring the sector's sustainability credentials? At least three roles are emerging:

Table 8.1 Global resource use and pollution loading attributable to buildings

Resources used in buildings	%
Energy	45–50
Water	50
Materials (by bulk)	60
Agricultural land loss to buildings	80
Timber	60 (90% of hardwood)
Coral reef	50 (indirect)
Rainforest	25 (indirect)
Pollution attributable to buildings	%
Air quality in cities	23
Global warming gases	50
Drinking-water pollution	40
Landfill waste	50
Ozone thinning	50

Source: Edwards (2010).

- The quantity surveyor's role in building sustainability assessment, especially with respect to optioneering – developing a whole-life value approach to meaningfully measure and assess building sustainability (see Chapter 3).
- Sustainability performance, zero carbon and property value – both in new and existing stock.
- Achieving sustainability value in construction procurement, especially in the public sector.

Although building sustainability and its assessment is more than fifteen years old, the current regime of assessment fails to offer the desired coverage of all performance issues across the whole life of the built asset in line with the principles of sustainable development. Furthermore, current sustainability assessment tools reflect a culture of compliance with associated performance labels intending to reward best practice, but failing to inspire clients to tackle sustainability in an aspirational manner with a view to raising the bar (Bioregional, 2008). What quantity surveyors can and must do to such aspirations is to develop an assessment framework that is compatible with more advanced articulations of sustainability value, especially with respect to realistic assessment of options, helping clients make informed decisions based on sustainability value rather than mere economic value.

A critical part of the sustainability debate is the more pressing need to stabilise atmospheric carbon concentrations in order to arrest the warming trend. Towards this end, many countries – including the UK – are aiming to decouple their economies from carbon-intensive sources of energy. While cleaning up energy production is the principal focus of such decarbonising attempts, demand reduction is equally key to a low/zero carbon future and the built environment has an important role to play in this regard. Quantity surveyors have a role in assessing the value dimensions and optioneering the most efficient route to low/zero carbon buildings.

A third unmet need in the creation of sustainable built environment is to embed sustainability value in the procurement of built assets. A vast majority of building sustainability assessment regimes are confined to assessing the environmental performance of buildings and fail to address their impact upon quality-of-life and the interrelationship between the two. Even within the current environmental focus, the emphasis is often on a narrow range of technical issues such as energy performance (e.g. *The Energy Performance in Buildings Directive* – European Commission, 2002) and material use (Environmental Product Declaration [EPD], 2008). Sustainability in its widest sense, delivering quality of life while improving environmental performance, is yet to be put into operation. Existing frameworks such as the BREEAM (BRE, 2009) and LEnSE system (LENSE Partners, 2007) remain uncoupled from the procurement side of the building's life cycle and neither account for the effect of procurement on sustainability nor explicitly state the procurement conditionalities. There is a great need for the core skills of a quantity surveyor to have a bearing here. As many of the decisions that ultimately influence building sustainability performance are made early on in the project, the building's design, construction, operation and procurement processes need to be assessed for whole-life sustainability in ways that facilitate early optioneering (European Commission, 2007b). In addition, in order to assist comparability and increase transparency in decision-making, greater standardisation of building assessment methodologies is necessary. Given the enormous potential for sustainability improvement in the construction industry and in the associated assessment methodologies, it is also necessary to recognise that sustainable building practices are necessarily path-breaking (cf. European Commission, 2007b). It is therefore necessary to value the nature of innovation and the amount of risk-taking involved in creating a new paradigm of building delivery, and these need to be incorporated into the determination of the Economically Most Advantageous Tender (EMAT) model of procurement.

Quantity surveyors have a specific role to play in the procurement of sustainable public buildings. Publicly procured exemplar sustainable buildings offer possibilities to drive the step-change needed for built environment sustainability. Forty per cent of the construction demand comes from the public sector.

Assessment of sustainability of the built environment – state of the art and current gaps in knowledge

Publicly procured construction projects are increasingly demanding sustainability assessment as a component of planning, client and funding requirements, representing a legislative and culture shift in procurement practice (European Commission, 2004a; Cole, 2005a; Lutzkendorf and Lornez, 2006). Yet the lack of integration of the three dimensions of sustainability (i.e. environmental, economic and social) and the near total failure to link assessment with decisions made across the project life cycle make true sustainability difficult to achieve (Walton *et al.*, 2005; Thomson *et al.*, 2009). Current assessment protocols offer virtually no feedback to contribute to better decision-making in the future (Brandon and Lombardi, 2005; Kaatz *et al.*, 2006; Thomson *et al.*, 2008). Such shortcomings call for the embedding of a whole life value approach to decision-making (ECCJ, 2007; Kelly *et al.*, 2004). It is also necessary to align assessment systems to focus on the ability to promote sustainable user behaviour (Cole, 2006; Head, 2008).

State of the art: assessment systems

For the built environment Walton *et al.* (2005) identified over 600 tools directly and indirectly relevant to sustainability assessment, and the mapping of this large landscape of tools has been the focus of a number of important projects (BEQUEST, 2000; Deakin *et al.*, 2001; CRISP, 2005; IEA ANNEX31, 2005; BTP, 2007; RICS, 2007). In a study for the US General Services Administration (GSA) sustainable building procurement process, Fowler and Rauch (2006) give an overview of more than a hundred tools, filtering them down to about thirty that are suitable for whole-building assessment, five of which are further analysed as appropriate for publicly procured non-residential buildings.

The focus of the vast majority of building sustainability assessment tools has been on the environmental performance (commonly in relation to biophysical inputs and outputs as well as impacts on local ecological health and on local and indoor environmental quality). These include:

1. Performance modelling software (e.g. for energy: EnergyPlus, DOE-2, BLAST, ESP-r, EcoTect, TAS, TRNSYS and many others (see BTP, 2007; Wong *et al.*, 2008).
2. Life-cycle assessment (LCA) tools (e.g. ATHENA, see ATHENA, 2008; TEAM, see TEAM, 2008; ENVEST, see ENVEST, 2008).
3. Whole-building rating systems (e.g. BREEAM, see BREEAM, 2007; LEED, see LEED, 2008; SBTool, see SBTool, 2007).
4. Environmental guidelines and checklists (e.g. Green Guide, see Green Guide, 2008; WBDG, see WBDG, 2008);
5. Product declarations, certifications and Labels (e.g. LENSE, see LENSE Partners, 2007; EPD, see EPD, 2008; Green-IT, see Green-IT, 2008).

Building assessment tools can be contrasted in a number of other important aspects:

1. Whether they assess new or existing buildings.
2. The different types of building they assess (e.g. schools, offices, homes, new-build, existing build).
3. The life cycle stages they assess (e.g. design, construction, operation, maintenance, decommissioning).
4. Their geographical relevance (reflected in their underlying databases).
5. The forms of their input and output (e.g. qualitative, quantitative, single value, range).
6. The media through which they are delivered (e.g. software or print).
7. Their intended user (e.g. architect, engineer, dedicated assessor) (see Forsberg and von Malmborg, 2004; Peuportier and Putzeys, 2005; Zhenhong *et al.*, 2006; Haapio and Viitaniemi, 2008; Chew and Das, 2008; Ding, 2008).

However, despite the wide range of tools available, the actual number of assessed buildings remains relatively low – indicating the need for continued reflection on the role of assessment tools (Cole, 1999, 2005a) as well as to identify and realise opportunities for their improvement (ISO, 2006a; PRESCO, 2008).

State of the art: public procurement for value and sustainability

Public procurement for whole-life value

Public sector procurement represents a significant share of the world's GDP, with a 6 to 10 per cent average nationally (excluding compensation to public employees) (United Nations, 2008). Expenditure within the EU Member States is higher, with 16 per cent of GDP on average spent through this route, exerting significant influence over the market and direction of the overall economy (Steurer *et al.*, 2007). Around 40 per cent is represented by public sector projects (European Commission, 2007c). Several procurement routes are applicable within construction projects; traditional procurement (where the design process is separated from construction, with the completed design put out for competitive/negotiated tender), design and build, management contracting, construction management, framework agreements, two-stage tendering and private finance initiative (PFI)/public/private partnership (PPP) (OGC, 2007a). For example, in the UK, national government policy since 2000 is that publicly procured projects are required to follow one of three recommended routes: PFI (for those above £20 million), prime contracting and design and build; with traditional routes only being considered if they demonstrate more value than these (OGC, 2007a).

Given its significance, the selection of an appropriate procurement route for the delivery of such projects not only has a bearing on the value for money to the taxpayer, but it plays a major role in the overall sustainability of the national economy. Traditionally, time, cost and quality are the three main factors governing the selection of the procurement route and form the criteria against which significant improvement in performance is judged (Harris *et al.*, 2006). However, procurement decisions are increasingly based on an agenda of value for money over the life cycle and not on the initial capital cost alone, in addition to a need to demonstrate efficient value in its wider sense (Kelly *et al.*, 2004). The most appropriate life-cycle solutions to construction projects are optimised through value management (VM) (see Male *et al.*, 2007) during the procurement phase of the project. It is a principle of VM that ´the cost effectiveness equation takes account of the whole value of the project to the client´ (Kelly *et al.*, 2004; Kelly, 2007).

The delivery of sustainable public procurement (SPP) in the UK is recognised by the UK Sustainable Development Strategy (2005) through public sector policies and decisions to stimulate a shift in the market towards sustainable goods and services, in addition to providing the necessary leadership to negotiate a sustainable path through its consumption patterns to business and consumers. SPP within construction increasingly plays a significant role in maximising value, realising cost savings and reducing risk (Dickinson *et al.*, 2008) due to its focus on reducing resource consumption and by considering decisions against the principles of sustainable development across the stages of the project's life cycle. The policy context is the requirement that decisions should reflect long-term value for public expenditure as opposed to simple considerations of cost, time and quality. This provides a key link between achieving best value and the need to deliver this through a holistic, long-term perspective that is aligned to the principles of sustainable development (SPTF, 2006).

It may therefore be argued that long-term contracts such as PFI promote whole-life value and aid its management due to the assignment of the service delivery responsibility to the same private sector agent, usually a consortium of companies (Nisar, 2007). In the EU, the trend is towards the adoption of procurement routes that deliver the integration of design, construction, operation and ongoing maintenance functions through consideration of the whole-life value of the service or facility (ECCJ, 2007). These principles are reflected through the EU Public Procurement Directives (Directives 2004/17 and 2004/18 focusing on contracting organisations and specialists) and are being applied at a national level on an individual basis.

Green public procurement and sustainable public procurement

Since its recognised importance during the World Summit on Sustainable Development in Johannesburg in 2002 (WSSD, 2002) public procurement is increasingly conceived as a key driver for improving the environmental performance of goods and services (Weiss and Thurbon, 2006); a sentiment recently reiterated by the OECD Council (United Nations, 2008). This has prompted efforts across the globe to integrate environmental and wider sustainability considerations within public procurement (Bouwer *et al.*, 2005; Everett and Hoekman, 2005; Green *et al.*, 2005; DEFRA, 2006; Brammer and Walker, 2007; ECCJ, 2007; Forum for the Future, 2007; Unge *et al.*, 2007; MTF, 2008). Such efforts are seen as a test of ability and desire to 'walk the talk' and lead by example when it comes to making decisions that align with the principles of sustainable development (Steurer *et al.*, 2007). For example, in 2003, the European Commission called for member states to increase the level of green public procurement (GPP) and to elaborate national action plans to set targets and outline measures for implementing policy through the adoption of a communication on Integrated Product Policy (IPP) (European Commission, 2003). Although the EU Public Procurement Directives did not prescribe SPP, they did open up the possibility to consider social and/or environmental issues at an early stage of the procurement process through the incorporation of the principles of sustainable development (McCrudden, 2007).

A survey of EU member states in May 2007 revealed that nine out of the twenty-seven member states had adopted national SPP or GPP Action Plans, five drafted but not approved, two in preparation, one where the status was unclear, three had none, with seven not identified/unclear (Steurer and Konrad, 2007). The most advanced EU states are the so-called 'Green 7' – Austria, Denmark, Finland, Germany, Netherlands, Sweden and the UK – demonstrating 40 to 70 per cent of all tenders incorporating some environmental criteria, with the other twenty falling below 30 per cent (especially in the Central/Eastern European nations) (Steurer *et al.*, 2007). Currently, wide variation exists in the approach to SPP, both in national policy and practical application. For example, the UK has a comprehensive Strategic Action Plan for SPP which aims to be among the leaders in Europe on sustainable procurement by 2009 (UK Government, 2005). However, even in the UK, effective and transparent guides to SPP are rare.

Gaps in the state of the art

A review of the state of the art indicates five key gaps in the current state of knowledge:

1. Integrating sustainability assessment with the decision-making process.
2. Broadening the scope from 'green' to sustainability assessment.
3. Moving beyond mere mitigation of damage.
4. A need for standardisation.
5. Integrating whole-life value within a sustainable procurement process.

Recognising the importance of processes

The use of sustainability assessment tools has the potential to raise awareness of relevant issues among the built environment professions and to promote social learning (Thomson *et al.*, 2009). Assessment can foster a shared understanding of the contextual requirements of the building project, including during early stages where the decisions that have the greatest influence over sustainability performance are taken (Lutzkendorf and Lorenz, 2006). However, the challenge for assessment tools is to provide valuable outputs in the absence of detailed design information relating to the current project at the early design stages (Ding, 2008). The assessment tools' need to operate under different conditions of complexity is in tension with their various required roles (for example, to provide an accurate assessment but also to communicate broad overall performance). Furthermore, it is critical that design teams learn from the actual performance of past projects, thus underlining the importance of assessment tools for auditing and inter-project knowledge capture (Thomson *et al.*, 2009).

Broadening scope and referencing environmental limits

If the current environmental (or 'green') building assessment tools are to become sustainable building assessment tools, they need to evolve in two key directions. First, in regard to the particular performance aspects assessed, with sustainability assessment examining quality of life as well as environmental aspects of building performance (McIntyre, 2006; Colantonio, 2007). Second, in regard to the levels against which environmental performance aspects are assessed; with sustainability assessment referencing building performance against ecological carrying capacities rather than the performance of other buildings (Cole, 2006, 2007).

In terms of the former, key building performance aspects influencing quality of life and requiring greater attention during assessments are: indoor air quality, accessibility, safety, user comfort, cultural value and user satisfaction (European Commission, 2004b; McIntyre, 2006). Priority environmental performance aspects include ability of the building to avoid environmental risk (e.g. flooding), building functionality and serviceability, identifying opportunities for closed material loops, and design for deconstruction (European Commission, 2004b; Webster, 2007). For existing buildings, key assessment priorities include the assessment of embodied energy and materials, the building's useful lifespan, and issues relating to hazardous materials and contaminated surroundings (Lutzkendorf and Lorenz, 2006). Although this list is not exhaustive, Fenner and Ryce (2008) caution against introducing additional criteria within assessments: 'unmonitored increase in assessment focal points runs the risk of diffusing public awareness of, and professional interest in, improving the environmental performance of buildings' (Fenner and Ryce, 2008).

Future assessment tools need to do more than just add socio-economic criteria to the pot of environmental ones but be able to examine the interdependencies between ecological and quality-of-life requirements. However, the challenge of doing this is reflected in the long-running, and still unsettled debate on the use of absolute environmental limits during building sustainability assessment (Cooper, 1999; Rees, 1999; Lowe, 2006; Cole, 2007; Zimmerman and Kibert, 2007; Gasparatos *et al.*, 2008). Authors argue that building assessment tools need greater integration with life-cycle assessment (LCA) and that environmental impacts over the lifetime of the building (including its materials) should ultimately be referenced to the ability of nature to support these impacts when considered in aggregate with all the other environmental impacts. It is argued that only then would an assessment truly examine the contribution that the building makes to sustainable development and thus constitute a genuine sustainability assessment. Assessment tools need to be compatible with a 'developmental approach' where awareness of global ecological limits sets the strategic direction within which decisions are made rather than such limits representing the standard against which every decision is assessed (Lowe, 2006).

Beyond mitigation: bio-mimicry and the sustaining building

Principles such as *One Planet Living* (OPL) (Desai and King, 2006) seek to enable people to live happy, healthy lives within our fair share of the Earth's resources. These principles have provided the framework guiding the design of a number of developments including the BedZED and Greater Middlehaven developments in the UK and the Mata de Sesimbra ecotourism project in Portugal (Bioregional, 2008; and see Figure 8.1). The OPL principles recognise that to deliver greater sustainability performance there is a need to adopt a more holistic and contextualised approach to design and assessment. This will mean consideration of not only the performance aspects associated with the physical development but also how the development will promote sustainable lifestyles during construction and operation (for example, through the sourcing of local and sustainable food for the construction team and end-users or through developing fair-trade partnerships with local and more distant deprived communities).

Another recent trend is the 'sustaining building' (Cole, 2005b; Kibert, 2007; Reed, 2007), one which practices 'bio-mimicry' (Guy and Moore, 2004; Head, 2008) through the promotion of human well-being in balance with the biosphere. Such buildings do more than mitigate environmental burdens; they are adaptive and restorative. They promote ecological healing in the surrounding bioregion (for example, through the use of cladding producing run-off which neutralises surrounding soil contamination or the use of green-roofs to help regulate micro-climate) as well as in larger natural systems (for example, through the use of bio-materials that absorb atmospheric carbon) (Reed, 2007). It seems obvious that new assessment tools will be needed to guide the design and examine the performance of these buildings for the 'ecological age' (Head, 2008).

Leadership, markets and standardisation

As public concern over the state of the environment rapidly grows, there is no doubt that the importance of acting, and being seen to act, in ways which contribute positively to sustainable development will also rapidly increase. In line with this view there is evidence that building sustainability assessment is increasingly being seen as a valuable tool beyond examining performance but in helping transform the market as well as in demonstrating commitment and leadership on sustainability (Cole, 2005a). Building sustainability assessment is increasingly being integrated into the market: through performance-labelling schemes aimed at a building's potential 'consumers'; as part of property valuations; as conditions on insurance or lending; as part of corporate sustainability reporting; or by guiding socially responsible investment (Lutzkendorf and Lorenz, 2006). At the same time, governments recognise the need to demonstrate leadership through greater sustainability

Figure 8.1 Mata de Sesimbra eco-tourism project – Portugal

Source: RICS Project Management Faculty.

in publicly funded projects, with associated tendering procedures being updated accordingly (European Commission, 2004a). A fast-growing issue for assessment tools is that their processes and output are both more transparent and are better comparable within and across nations (Dammann and Elle, 2006; Lutzkendorf and Lorenz, 2006). For example, across the EU greater standardisation has been sought with regard to particular assessment algorithms, assessment procedures, minimum required sets of performance indicators, performance declarations and the methodological frameworks used to guide tool development (e.g. CEN TC 350; see European Commission, 2004a; ISO, 2006a, 2006b, 2007).

Transparent and inclusive protocols to help realise whole-life value through SPP

Despite increasing efforts around the globe to promote sustainability through procurement, much of the focus to date has been on environmental concerns overlooking social impacts (Unge *et al.*, 2007). For construction, the focus has been on green public procurement (GPP) rather than sustainable public procurement (SPP), leading to suboptimal alignment with whole-life value (Dickinson *et al.*, 2008).

The lack of application of SPP is partly due to:

1. confusion with the GPP agenda
2. misconception that the drive for 'value for money' represents a barrier to sustainability
3. variation in its application at policy level
4. lack of innovative procurement tools and initiatives that incorporate sustainability
5. lack of understanding of the principles and process implications of its application (see Brammer and Walker, 2007; Forum for the Future, 2007; Steurer *et al.*, 2007; Unge *et al.*, 2007; Dickinson *et al.*, 2008).

The key to addressing these concerns is increased stakeholder involvement, education, training and the provision of practical aids and handbooks to aid the transfer of knowledge and stimulate the required learning (DEFRA, 2006). The OGC's Framework for Sustainable Construction Procurement represents a rare attempt to meet some of these requirements in the UK context (OGC, 2007a, 2007b).

Thus, a significant contribution would be made to the delivery of SPP for publicly funded construction projects if a widely accepted protocol was developed to guide associated decisions and promote transparency and inclusiveness in these decisions.

The role of quantity surveyors in sustainability assessment

Given the importance of sustainability assessment in buildings and the state of the art in assessment protocols, we can discern three areas where the core skills of a quantity surveyor can maximise whole-life sustainability of the construction sector: sustainability optioneering in design, achieving sustainability value in procurement (especially public procurement) and valuing sustainability in properties.

Sustainability optioneering in design

Sustainability optioneering (i.e. comparison of design options in light of sustainability value) need to be underpinned by a logical, transparent and evidence-based system of life-cycle cost assessment. The 'cost' in question ought to be informed by sustainability value, rather than by mere economic value. To help with such a 'sustainability value' Kelly and Hunter (2009) recently introduced a five-step process to life-cycle cost appraisal of a sustainability project, which is itself an improvement of the standard life-cycle assessment (LCA) methods:

Step 1 – Project identifiers

Identify and describe the project including the basis for the calculation (i.e. whether the data is parametric or obtained from manufacturers/suppliers), and the time zero point for all calculations. The type of life-cycle cost calculation, prediction of cash flow or option appraisal (with or without a base case) may be included in the general description. This identifies how the data will be used.

Step 2 – Study periods

Determine the length of the study period and also the unit of time. The units of time and the interest rate must correlate (i.e. if the unit of time is months then the interest rate must be a percentage rate per month). It may be advantageous to set up any model to calculate over a number of time periods so that options can be quickly compared rather than running repetitive sensitivity checks.

Step 3 – Inflation rate and discount rate

The inflation rate only is used when predicting a cashflow over time for the purposes of budgeting, cost planning, tendering, cost reconciliation and audit.

Discount rates are used when comparing two or more dissimilar options during an option appraisal exercise or when comparing tenders which have an FM constituent. The discount rate will be legislated, calculated or given by the client. Public sector option appraisal calculations tend to use the discount rate issued by HM Treasury (January 2008) which remains at 3.5 per cent. A calculated discount rate takes a relevant rate of interest (e.g. the bank rate), and adjusts this for inflation. A client-nominated discount rate is used when considering options against strict internal rate of return or opportunity cost of capital criteria.

Step 4 – Gather data

Data will be obtained from parametric sources (e.g. BCIS Running Costs Online), or from first principles either by calculation (e.g. energy calculation), or from manufacturers or suppliers. Data gathered from manufacturers or suppliers should include the detail illustrated above.

Step 5 – Model construction and analysis

There are few commercially available software packages which allow for the type of calculation described above. Many quantity surveying practices have a life-cycle cost package developed and used in-house. These are generally spreadsheet based. The illustration below was constructed using a spreadsheet (Kelly and Hunter, 2009).

Achieving sustainability value in construction procurement, especially public procurement

UK government policy since 2000 has stated that public projects should be procured by one of three recommended routes: PFI (for those above £20 million), prime contracting and design and build; with traditional routes only being considered if they demonstrate more value than these recommended routes (OGC, 2007a). The significance of supply chain management and the role played by partnering contracts are increasingly targeted as approaches to integrating life-cycle decision-making. Both the OGC in the UK and the EU have adopted procurement routes that deliver the integration of design, construction, operation and ongoing maintenance functions through the consideration of whole-life value of the service or facility (OGC [2007a] and the EU Public Procurement Directives – Directives 2004/17 and 2004/18).

A lack of understanding of the principles and process implications of applying SPP in practice is a significant barrier to the effective use of SPP. In this light the recent development of a sustainable procurement standard (BS 8903:2010, BSI, 2010) will go a long way towards clarifying the practical implementation of SPP in construction projects.

Given the lack of uniform standards, we present below the UK government's framework around which all publicly procured projects will be required to be managed (Government Construction Client's Panel, 2000):

- Reuse existing built assets – Consider the need for new build. Refurbishment/reuse may work better. Think brownfield wherever possible for new construction.
- Design for minimum waste – design outwaste both during construction and from the useful life – and afterlife – of the building or structure. Think whole-life costs. Involve the supply chain. Specify performance requirements taking care to encourage more efficient use of resources. Think about using recycled materials.

- Aim for lean construction – work on continuous improvement, waste elimination, strong user focus, value for money, high-quality management of projects and supply chains, improved communications.
- Minimise energy in construction – be aware of the energy consumed in the production and transport of construction products.
- Minimise energy in use – consider more energy-efficient solutions in design including passive systems using natural light, air movement and thermal mass, as well as solutions involving energy produced from renewable sources.
- Do not pollute – understand your environmental impacts and have policies and systems to manage them positively. Use environmental management systems under ISO 14001 or EMAS. Specify adoption of the Considerate Constructors Scheme or similar.
- Preserve and enhance biodiversity – look for opportunities throughout the construction process – from the extraction of raw materials, through the construction phase, to the landscaping of buildings and estates – to provide and protect habitats.
- Conserve water resources – design for increased water efficiency in building services and water conservation within the built environment.
- Respect people and their local environment – be responsive to the community in planning and undertaking construction. Consider all those who have an interest in the project (employees, the local community, contractors).
- Set targets – measure and compare your performance with others. Set targets for continuous improvement. Develop appropriate management systems.

A key to effective sustainable procurement is to integrate it with sustainability assessment. As sustainability assessment becomes not just a reactive tool that supports goal-orientated decision making, but a proactive tool that guides predominantly subjective decisions within building projects, its consideration within this context becomes increasingly relevant (Thomson *et al.*, 2008). If whole-life value is to evolve as a tool for aiding the delivery of SPP and therefore aid the delivery of sustainable building performance, sustainability assessment has a significant role to play in providing the necessary tools to inform value-based decisions in a holistic, robust, transparent and reflective manner.

Valuing sustainability in properties

There is some evidence to suggest that the property market does pay a small premium to buildings rated 'green'. In the US, green-rated office buildings fetch 3 per cent higher rental income per square metre than unrated buildings (Eichholtz *et al.*, 2009). *Energy star*-rated buildings yield 5.9 per cent higher net incomes per square foot (due to 9.8 per cent lower utility expenditures, 4.8 per cent higher rents and 0.9 per cent higher occupancy rates), 13.5 per cent higher market values per square foot, 0.5 per cent lower cap rates, and appreciation and total returns similar to other office properties (Pivo and Fisher, 2009). Yet the valuation of sustainability in property remains highly problematic, largely due to uncertainties as to what sustainable real estate is. Meins *et al.* (2010) identified three main challenges to valuation in general, but this applies to sustainable property valuation as well: how to deal with uncertainties (valuation uncertainty), lack of transparency (valuation black box), and the tendency of valuations to lag behind market trends (valuation lag). While sustainable building assessment systems can help in valuing sustainability, these systems rarely take into account economic sustainability and therefore do not directly help the valuation of properties.

Several recent attempts are beginning to emerge in relevant literature to overcome the above problem. The RICS's Valuation Information Paper (VIP) is a significant step in this regard. A further tool – the Economic Sustainability Indication (ESI) – was recently proposed by Meins *et al.* (2010). ESI measures the risk of property to lose value and the opportunity to gain value due to future developments (e.g. climate change or rising energy prices). Five groups of value-related sustainability features are identified:

1. flexibility and polyvalence
2. energy and water dependency
3. accessibility and mobility
4. security
5. health and comfort.

By minimising the risk of loss in value through future developments, those sustainability features contribute to the property value. Their effects on property value are quantified by risk modelling.

Table 8.2 shows the sustainability features of the building included in the ESI from a financial point of view

Refurbishment and whole-life sustainability

Rules, assessment protocols and strategies for sustainable buildings focus overwhelmingly on new stocks. However, in terms of the narrower legislatively mandated low/zero carbon targets we need to tackle the existing stock. The UK Climate Change Act of 2008 requires a carbon reduction of 60 per cent from the 1990 levels by 2050. However, over 80 per cent of the building stock that will be in use in 2050 has already been built. More than half of the European building stock is over forty years old and the annual replacement rate for non-residential buildings is a mere 1–1.5 per cent and 0.07 per cent for residential buildings (Barlow and Fiala, 2007; Poel *et al.*,

Table 8.2 A risk-based approach to sustainability valuation on properties – the ESI method

Sustainability features	*External conditions*
1. Flexibility and polyvalence 1.1 Flexibility of use 1.2 Adaptability to users	Demographics, structure of households
2. Energy and water dependency 2.1 Energy demand and production 2.2 Water use and wastewater disposal	Climate change, energy and water prices
3. Accessibility and mobility 3.1 Public transport 3.2 Pedestrians and non-motorized vehicles 3.3 Accessibility	Percentage of the aged population, cost of fuels
4. Safety and security 4.1 Location regarding natural hazards 4.2 Building safety and security measures	Climate change, need for safety and security
5. Health and comfort 5.1 Indoor air quality 5.2 Noise 5.3 Daylight 5.4 Radiation 5.5 Ecological construction materials	Need for safety, health awareness, building services

Source: Meins *et al.* (2010).

2007). Thus even if we start doing the right thing (i.e. build zero carbon buildings) today, the net effect on the carbon concentrations in the atmosphere will only be marginal at best.

The importance of existing building stock goes up not only because of their central role in decarbonising the economy but also due to the rising operational costs to meet increasingly stringent legislative environment and the sharply rising energy costs. For example, during 2002 and 2007 the costs for heating (oil) and electricity in Germany increased by 63.8 per cent and 26.3 per cent respectively (see GuG, 2008). Added to these, restrictions on credit availability to new-build developments in many leading economies offer a huge opportunity 'to take cost-effective measures and transform (existing stock) to resource-efficient and environmentally sound buildings, with an increased social and financial value' (Poel *et al.*, 2007, p. 394).

Although the opportunities are plenty, refurbishment of built assets has several barriers to overcome. Chau *et al.* (2003) identified three key barriers:

1. Fragmented ownership of multi-family houses and associated negotiation costs between owners which may be so high that it prohibits a collective decision from being made.

2. Inability to realise whether the benefits of refurbishment will outweigh the costs.
3. Lack of cost and quality information on contractors and their works.

While the surveying professions have little leeway with respect to item (1) above, it does have the ability to significantly influence the other two items. The nexus between lack of data and poor decision-making can only be broken by providing property market actors with appropriate feedback on both the environmental and social aspects of building performance as well as on their interrelations with financial performance and property value. In this respect, valuation professionals and the valuation process itself can and should play an important role as mainstream financial professionals are unwilling to include sustainability issues in property investment and financing decisions unless and until sustainable building features and related performance are integrated into property valuations; in other words, unless *the financial sector understands the benefits of green to the net value of an asset* (RICS, 2005, p. 17). In addition, the valuation professionals' central role as well as major responsibility is also due to their function as the independent pivotal point for all property-related information. They have the role of 'information managers' in a market where the distribution of information is traditionally considered asymmetrical (Lorenz *et al.*, 2008). Another barrier to the correct understanding of the value of existing buildings is the valuation professionals' inability to value existing assets inclusively. Valuation in the current sense is almost exclusively financial: no account of the intangible benefits of an existing built asset is included:

> An insensible 'intrusion of the financial method in the real estate field' fails taking into account that property assets do have major environmental and social impact with tangible consequences for our every-day life and well-being. Property assets have the capacity to improve the quality of space and life. Buildings, groups of buildings and the public space stand for something, have meaning, give identity, project image, shape perceptions, express territory (not always a good thing of course), promote social cohesion, etc. All this has indirect, but to a certain extent, also direct monetary benefits, which are disregarded in current practice. Consequently, whenever financial methods and techniques are applied without taking into account the specific nature of property assets and investments and without prior adjustment to the subject matter of investigation and inquiry, the advice given on that basis is likely to be misleading.
>
> (Lorenz *et al.*, 2008)

Bibliography

ATHENA (2008). *ATHENA Impact Estimator for Buildings*, Merrickville, ON, Canada: ATHENA Sustainable Materials Institute. Available online at http://www.athenasmi.org/tools/impactEstimator/ (accessed 9 November 2010).

Barlow, S and Fiala, D. (2007). Occupant comfort in UK offices – How adaptive comfort theories might influence future low energy office refurbishment strategies, *Energy and Buildings*, 39, pp. 837–846.

BEQUEST (2000). *Building Environmental Quality for Sustainability Through Time*, project website: http://research.scpm.salford.ac.uk/bqextra/ (accessed 28 December 2008).

Bioregional (2008). Bioregional website: http://www.bioregional.com/ (accessed 28 December 2008).

Bouwer, M., de Jong, K., Jonk, M., Berman, T., Bersani, R., Lusser, H., Nissinen, A., Parikka, K. and Szuppinger, P. (2005). *Green Public Procurement in Europe 2005 – Status Overview*, Korte Spaarne, AJ Haarlem, the Netherlands at http://europa.eu.int/comm/environment/gpp/media.htm#state.

Brammer, S. and Walker, H. (2007). *Sustainable Procurement Practice in the Public Sector: An International Comparative Study*, University of Bath, School of Management Working Papers Series, 16.

Brandon, P. and Lombardi, P. (2005). *Evaluating Sustainable Development in the Built Environment*, Oxford: Blackwell.

BRE (2009). GreenPrint – enabling sustainable communities, Building Research Establishment, website: http://www.bre.co.uk/page.jsp?id=1290 (last accessed 6 January 2009).

BREEAM (2007). *Building Research Establishment Environmental Assessment Method*, tool website: www.breeam.org (accessed 31 December 2008).

British Standards Institution (BSI) (2010). *BS 8903:2010, Principles and Framework for Procuring Sustainably* London: BSI

Brunklaus, B., Thormark, C. and Baumann, H. (2010). Illustrating limitations of energy studies of buildings with LCA and actor analysis, *Building Research and Information*, 38, 3, pp. 265–279.

BTP (2007). Building energy software tools directory. Building Technologies Program of the US Department of Energy, project website: http://apps1.eere.energy.gov/buildings/tools_directory/ (accessed 30 December 2008).

Chau, K.W., Leung, A.Y.T., Yiu, C.Y. and Wong, S.K. (2003). Estimating the value enhancement effects of refurbishment, *Facilities*, 21, 1/2, pp. 13–19.

Chew, M.Y.L. and, Das, S. (2008). Building grading systems: a review of the state-of-the-art, *Architectural Science Review*, 51, pp. 3–13.

Colantonio, A. (2007). *Social sustainability: An exploratory analysis of its definition, assessment methods, metrics and tools*, EIBURS Working Paper Series.

Cole, R. (1999). Building environmental assessment methods: clarifying intensions, *Building Research and Information*, 27, pp. 230–246.

Cole, R. (2005a). Building environmental assessment methods: redefining intentions and roles, *Building Research and Information*, 33, pp. 455–467.

Cole, R. (2005b). Building green: moving beyond regulations and voluntary initiatives, *Policy Options*, July–August.

Cole, R. (2006). Building environmental assessment: changing the culture of practice, *Building Research and Information*, 34, pp. 303–307.

Cole, R (2007). Reframing environmental performance goals for buildings, Proceedings, *International Conference on Whole Life Urban Sustainability and its Assessment*, Glasgow Caledonian University, Glasgow, ISBN- 13 978–1–905866–13–7; ISBN- 10 1–905866–13–5.

Cooper, I. (1999). Which focus for building assessment methods-environmental performance or sustainability?, *Building Research and Information*, 27, pp. 321–331.

CRISP (2005). A European thematic network on construction and city related sustainability indicators, project website: http://crisp.cstb.fr (accessed December 2008).

Dammann, S. and Elle, M. (2006). Environmental indicators: establishing a common language for green building, *Building Research and Information*, 34, pp. 387–404.

Deakin, M., Curwell, S. and Lombardi, P. (2001). BEQUEST: The framework and directory of assessment methods, *International Journal of Life Cycle Assessment*, 6, pp. 373–383.

DEFRA (2006). *Procuring the Future - The Sustainable Procurement Task Force, National Action Plan*, Department for Environment, Food and Rural Affairs, London, website: http://www.defra.gov.uk/sustainable/government/documents/full-document.pdf (accessed 06 February 2011).

Desai, P. and King, P. (2006). *One Planet Living – A Guide to Enjoying Life on Our One Planet*, Bristol: Alastair Sawday Publishing.

Dickinson, M., McDermott, P. and Platten, A. (2008). Implementation of sustainable procurement policy innovations, CIB W107 Construction in Developing Countries International Symposium, *Construction in Developing Countries: Procurement, Ethics and Technology*, 16–18 January, Trinidad and Tobago, West Indies.

Ding, G.K.C. (2008). Sustainable construction – the role of environmental assessment tools, *Journal of Environmental Management*, 86, pp. 451–464.

ECCJ (2007). *Sustainable Procurement in the European Union Proposals and Recommendations to the European Commission and the European Parliament. European Coalition for Corporate Justice*, Brussels: Commission of the European Communities.

Edwards, B. (2010). *Rough Guide to Sustainability: A Design Primer*, London: RIBA Publishing.

Eichholtz, P.M.A., Kok. N. and Quigley, J.M. (2010). Doing well by doing good? Green office buildings, *American Economic Review*, 100, pp. 2492–2509.

ENVEST (2008). *ENVEST*, tool website: http://envestv2.bre.co.uk/ (accessed 31 December 2008).

EPD (2008). Environmental Product Declaration, 89/106/EEC website: www.environdec.com (accessed 31 December 2008).

European Commission (2002). *Directive on Energy Performance of Buildings*, 2002/91/EC, Brussels: Commission of the European Communities.

European Commission (2003). *Integrated Product Policy, building on environmental lifecycle thinking*, Communication from the Commission to the Council and the Parliament of the European Union, Commission of the European Communities, COM (2003) 302 final, Brussels: Commission of the European Communities.

European Commission (2004a). *Mandate on the Development of Horizontal Standardised Methods for the Assessment of the Integrated Environmental Performance of Buildings,* European Commission's Standardisation Mandate to CEN M/ 350, 2004, Brussels: European Commission.

European Commission (2004b). *Towards a Thematic Strategy on the Urban Environment.* COM(2004)60, Brussels: European Commission.

European Commission (2006). *Buying Green– Handbook on Green Public Procurement*, Office for Official Publications of the European Communities, Luxemburg. Available online at http://ec.europa.eu/environment/gpp/pdf/int.pdf (accessed 11 December 2008).

European Commission (2007a). *A Lead Market Initiative for Europe*, Communication from the Commission to the Council, The European Parliament, The European Economic and Social Committee and the Commission of the Regions, COM (2007) 860, Brussels: Commission of the European Communities.

European Commission (2007b). *A Lead Market Initiative for Europe: Action Plan for Sustainable Construction*, Communication from the Commission to the Council, The European Parliament, The European Economic and Social Committee and the Commission of the Regions, SEC (2007) 1729, Brussels: Commission of the European Communities.

European Commission (2007c). *A Lead Market Initiative for Europe: Action Plan for Sustainable Construction*, Communication from the Commission to the Council, The European Parliament, The European Economic and Social Committee and the Commission of the Regions, SEC (2007) 1729, Brussels: Commission of the European Communities.

Everett, S. and Hoekman, B. (2005). Government procurement: market access, transparency, and multilateral trade rules, *European Journal of Political Economy*, 21, pp. 163–183.

Fenner, R.A. and Ryce T. (2008). A comparative analysis of two building rating systems. Part 1: Evaluation, *Engineering Sustainability*, March 2008, ES1, pp. 55–63.

Forsberg, A. and von Malmborg F. (2004). Tools for environmental assessment of the built environment, *Building and Environment*, 39, pp. 223–228.

Forum for the Future (2007). *Buying a Better World: Sustainable Public Procurement*, London: Forum for the Future.

Fowler, K.M. and Rauch, E.M. (2006). *Sustainable Building Rating Systems Summary*, Pacific Northwest National Laboratory. Available online at https://www.usgbc.org/ShowFile.aspx?DocumentID=1915 (accessed 16 December 2008).

Gasparatos, A., El-Haram, M. and Horner, M. (2008). A critical review of reductionist approaches for assessing the progress towards sustainability, *Environmental Impact Assessment Review*, 28, pp. 286–311.

Government Construction Client's Panel (GCCP) (2000). *Achieving Sustainability in Construction Procurement*, London: Office of Government Commerce.

Green, S., Fernie, S. and Weller, S. (2005). Making sense of supply chain management: a comparative study of aerospace and construction, *Construction Management and Economics*, 23, pp. 579–593.

Green Guide (2008). *Green Guide to Specification*, tool website. Available online at www.thegreenguide.org.uk (accessed 31 December 2008).

Green-I.T. (2008). Implementation status of energy performance EU standards in the building construction industry. Available online at www.green-it.eu/wacom.aspx?idarchitecture=21&Country= (accessed 31 December 2008).

GuG (2008). Entwicklung der Wohnnebenkosten in Deutschland, GuG-Grundstücksmarkt und Grundstückswert, *GuG Aktuell*, Ausgabe 1, S. 3–4 (quoted in Lorenz *et al.*, 2008).

Guy, S. and Moore, S.A. (2004). The paradoxes of sustainable architecture, in S. Guy and S.A. Moore (eds) *Sustainable Architectures: Cultures and Natures in Europe and North America*, London: E&FN Spon.

Haapio, A. and Viitaniemi, P. (2008). A critical review of building environmental assessment tools, *Environmental Impact Assessment Review*, 28, pp. 469–482.

Harris, F., McCaffer, R. and Edum-Fotwe, F. (2006). *Modern Construction Management*, Oxford: Blackwell.

Head, P. (2008). Entering the Ecological Age, The Institution of Civil Engineers Brunel Lecture Series, London: Institution of Civil Engineers.

IEA ANNEX31 (2005). Energy related environmental impact of buildings, project website. Available online at http://www.greenbuilding.ca/annex31/ (accessed December 2008).

Innovation and Growth Team (IGT) (2010). *Low Carbon Construction: Emerging Findings*, BIS/17/03/10/NP. URN 10/671, London: Department for Business Innovation and Skills.

ISO (2006a). *Sustainability in Building Construction – Framework for Methods of Assessment for Environmental Performance of Construction Works – Part 1: Buildings*, ISO/TS 21931-1: 2006.

ISO (2006b). *Sustainability in Building Construction – Sustainability Indicators – Part 1: Framework for Development of Indicators for Buildings*, ISO/TS 21929-1: 2006.

ISO (2007). *Sustainability in Building Construction – Environmental Declaration of Building Products*, ISO 21930:2007.

Kaatz, E., Root, D., Bowen, P. and Hill, R. (2006). Advancing key outcomes of sustainability building assessment, *Building Research and Information*, 34, pp. 308–320.

Kelly, J. (2007). Making client values explicit in value management workshops, *Construction Management and Economics*, 25, pp. 435–442.

Kelly, J. and Hunter, K. (2009). *Life Cycle Costing of Sustainable Design*, London: RICS. Available online at www.rics,org (accessed 26 April 2010).

Kelly, J., Male, S. and Graham, D. (2004). *Value Management of Construction Projects*, Oxford: Blackwell.

Kibert, C.J. (2007). The next generation of sustainable construction, *Building Research and Information*, 35, pp. 595–601.

LEED (2008). *Leadership in Energy and Environmental Design*, tool website. Available online at www.usgbc.org/LEED (accessed 31 December 2008).

LENSE Partners (2007). *Development of a Sustainability Assessment Methodology: Framework and Content*. Available online at http://www.lensebuildings.com/downloads/LEnSE_Stepping Stone 2 LR.pdf (accessed 29 December 2008).

Lorenz, D., d'Amato, M., Des Rosiers, F., Elder, B., van Genne, F., Hartenberger, U., Hill, S., Jones, K., Kauko, T., Kimmet, P., Lorch, R., Lützkendorf, L., and Percy, J. (2008). *Sustainable Property Investment and Management: Key Issues and Major*

Challenges, London: RICS. Available online at www.rics.org (accessed 23 April 2010).

Lowe, R. (2006). Defining absolute environmental limits for the built environment, *Building Research and Information*, 34, pp. 405–415.

Lutzkendorf, T. and Lorenz, D. (2006). Using an integrated performance approach in building assessment tools, *Building Research and Information*, 34, pp. 334–356.

Male, S., Kelly, J., Gronqvist, M. and Graham, D. (2007). Managing value as a management style for projects, *International Journal of project management*, 25, pp. 107–114.

McCrudden, C. (2007). *Buying Social Justice: Equality, Government Procurement, and Legal Change*, Oxford: Oxford University Press.

McIntyre, M.H. (2006). *A Literature Review of the Social, Economic and Environmental Impact of Architecture and Design*, Edinburgh: Scottish Executive.

Meins, E,, Wallbaum, H., Hardziewski, R. and Feige, A. (2010). Sustainability and property valuation: a risk-based approach, *Building Research and Information*, 38, pp. 280–300.

Monk, S., Grant, F.L., Markkanen, S. and Jones, M. (2010). *UK Government National Carbon Reduction Targets and Regional Housing Market Dynamics: Compatible or Contradictory?*, London: RICS.

MTF (2008). *Marrakech Task Force on Sustainable Public Procurement (MTF on SPP)*, Federal Office for the Environment (FOEN). Available online at http://www.unep.fr/scp/marrakech/tasforces/pdf/procurement2.pdf (accessed 06 February 2011).

Nisar, T.M. (2007). Value for money drivers in public private partners schemes, *International Journal of Public Sector Management*, 20, pp. 147–156.

Organisation for Economic Co-operation and Development (OECD) (2003). *Environmentally Sustainable Buildings: Challenges and Policies*, Paris: OECD.

OECD (2008). *Report on Workshop on Sustainability Assessment Methodologies*, Amsterdam: OECD.

OGC (2007a). *Procurement and Contract Strategies– Achieving Excellence in Construction Procurement Guide*, London: Office of Government Commerce, HM UK Government.

OGC (2007b). *Sustainability: Achieving Excellence in Construction Procurement Guide*, London: Office of Government Commerce, HM UK Government.

Peuportier, B. and Putzeys, K. (2005). *Inter-comparison and Benchmarking of LCA-based Environmental Assessment and Design Tools – Final Report*, European Thematic Network on Practical Recommendations for Sustainable Construction (PRESCO).

Pivo, G. and Fisher, J.D. (2009). Investment returns from responsible property investments: energy efficient, transit-oriented and urban regeneration office properties in the US from 1998–2008. Working Paper, Responsible Property Investing Center, Boston College/University of Arizona/Benecki Center for Real Estate Studies, Indiana University, Boston, MA/Tucson, AZ/Bloomington, IN.

Poel, B., van Cruchten, G. and Balaras, C.A. (2007). Energy performance assessment of existing dwellings, *Energy and Buildings*, 39, pp. 393–403.

PRESCO (2008). *European Thematic Network on Practical Recommendations for Sustainable Construction*, project website. Available online at http://www.etnpresco.net/generalinfo/index.html (accessed December 2008).

Reed, B. (2007). Shifting from sustainability to regeneration, *Building Research and Information*, 35, pp. 674–680.

Rees, W. (1999). The built environment and the ecosphere: a global perspective, *Building Research and Information*, 27, pp. 206–220.

RICS (2005). *Green Value – Green Buildings, Growing Assets*, published by the Royal Institution of Chartered Surveyors. Available online at www.rics.org/Practiceareas/Property/Green+value.htm (accessed 18 November 2007).

RICS (2007). *A Green Profession? RICS Members and the Sustainability Agenda, Research Report*, London: The Royal Institution of Chartered Surveyors.

RICS (2009). *Valuation Information Paper 13 – Sustainability and Commercial Property Valuation*, London: The Royal Institution of Chartered Surveyors.

SBTool (2007). *Sustainable Building Tool*, tool website. Available online at http://www.iisbe.org/iisbe/sbc2k8/sbc2k8-dwn.htm (accessed 31 December 2008).

Steurer, R. and Konrad, A. (2007). *Sustainability Public Procurement Initiatives in EU Member States: Preliminary Summary of the Survey Results*, Report for European Commission- Employment, Social Affairs and Equal Opportunities.

Steurer, R., Berger, G., Konrad, A. and Martinuzzi, A. (2007). *Sustainable Public Procurement in EU Member States: Overview of Government Initiatives and Selected Cases– Final Report to the EU High-Level Group on CSR*, Report for European Commission- Employment, Social Affairs and Equal Opportunities DG (Tender No. VT/2005/063).

Sustainable Procurement Task Force (SPTF) (2006). *Procuring the Future: Sustainable Procurement National Action Plan*, London: SPTF, Dept for Environment, Food and Rural Affairs (DEFRA). Available online at http://www.defra.gov.uk/sustainable/government/documents/full-document.pdf (accessed 9 November 2010).

TEAM (2008). *Tool for Environmental Analysis and Management*, tool website. Available online at http://www.ecobilan.com/uk_team.php (accessed 31 December 2008).

Thomson, C.S., El-Haram, M.A. and Hardcastle, C. (2009). Managing knowledge of urban sustainability assessment, *Proceedings of the ICE Engineering Sustainability*, 162, pp. 35–43.

Thomson, C.S., El-Haram, M.A., Hardcastle, C. and Horner, R.M.W. (2008). Developing an urban sustainability assessment protocol reflecting the project lifecycle, in A. Dainty (ed.) *Proceedings of the 24th Annual ARCOM Conference*, 1–3 September, Cardiff: Association of Researchers in Construction Management, 1155–1164.

UK Government (2005). *Securing the Future: UK Sustainable Development Strategy*, London: HMSO.

Unge, M., Ryding, S.O. and Frenander, C. (2007). *Environmental Management Systems and Environmental Declarations in Purchasing and Procurement*, Swedish Environmental Management Report, 2007: 2.

United Nations (2008). *Public Procurement as a Tool for Promoting more Sustainable Consumption and Production Patterns*, Sustainable Development Innovation Briefs, Issue 5, August.

Walton, J.S., El-Haram, M., Castello, H., Horner, R.M.W., Price, A.D.F. and Hardcastle, C. (2005). Integrated assessment of urban sustainability, *Engineering Sustainability*, 158 (ES2), pp. 57–65.

WBDG (2008). *Whole Building Design Guide*, tool website. Available online at http://www.wbdg.org/ (accessed 31 December 2008).

Webster, M.D. (2007). Structural Design for Adaptability and Deconstruction: A Strategy for Closing the Materials Loop and Increasing Building Value, *New Horizons and Better Practices*, Proceedings of Sessions of the 2007 Structures Congress, Long Beach, California, USA, 16–19 May.

Weiss, L. and Thurbon, E. (2006). The business of buying American: public procurement as trade strategy in the USA, *Review of International Political Economy*, 13, pp. 701–724.

Wong, I.L., Eames, P.C. and Singh, H. (2008). Predictions of heating and cooling energy demands for different building variants with conventional building fabric and selected fabric interventions, *Solpol 2008*, Warsaw, Poland, 22–26 September.

WSSD (2002). *Plan of Implementation of the World Summit on Sustainable Development*, New York: United Nations.

Zhenhong, G., Wennersten, R. and Getachew, A. (2006). Analysis of the most widely used building environmental assessment methods, *Environmental Sciences*, 3, pp. 175–192.

Zimmerman, A. and Kibert, C.J. (2007). Informing LEED's next generation with The Natural Step, *Building Research and Information*, 35, pp. 681–689.

Index